T0340039

Pi (π) in Nature, Art, and Culture

Studies in Mathematics
in the Arts and Humanities

VOLUME 1

Pi (π) in Nature, Art, and Culture

Geometry as a Hermeneutic Science

By

Marcel Danesi

BRILL

LEIDEN | BOSTON

Cover illustration: "Mathematics Multiplication Table Pi Fractal" - clipart, KISSCCo,
https://www.kisscco.com/clipart/mathematics-multiplication-table-pi-fractal-pi-fra-8ihrsg/,
Creative Commons Zero (CC0) license.

The Library of Congress Cataloging-in-Publication Data is available online at http://catalog.loc.gov

Typeface for the Latin, Greek, and Cyrillic scripts: "Brill". See and download: brill.com/brill-typeface.

ISSN 2666-2299
ISBN 978-90-04-43337-3 (hardback)
ISBN 978-90-04-43339-7 (e-book)

Contents

Preface

Don't become a mere recorder of facts, but try to penetrate the mystery of their origin.

IVAN PAVLOV (1849–1936)

∴

The nineteenth-century American essayist and poet, Henry David Thoreau, wrote the following discerning words in *The Maine Woods* about humanity's deep-rooted need to understand the mysteries of existence (Thoreau 1864: 36):

> Talk of mysteries! Think of our life in nature, daily to be shown matter, to come in contact with it, rocks, trees, wind on our cheeks! The *solid* earth! The *actual* world! The *common sense*! *Contact! Contact! Who* are we? *Where* are we?

The word *mystery* comes from the ancient Greek *mustērion*, designating a "sacred secret." It was applied to characterize the ancient polytheistic cults (Hall 1973, Mishlove 1993, Bremmer 2014), of which the one founded by the Greek mathematician and philosopher Pythagoras (c. 580–500 BCE) was the most famous.

The central objective of the Pythagorean cult was to investigate mathematics as a code that would help them unlock the secrets of the universe. Pythagoras had discovered that plucking strings or striking objects in a certain way produced harmonious sounds that could be described with specific numerical ratios. From this, he reasoned that the planets moving in orbits also produced them, being part of the same physical world—a view that musical harmonies informed the structure of the universe, which came to be known as the theory of *musica universalis* (literally, universal music), or the "music of the spheres." The harmonious sounds emitted by the planetary orbital revolutions are imperceptible to the human ear, unlike human musical sounds which are audible. Legend has it that Pythagoras could actually hear the celestial sounds—a gift conferred on him by the Egyptian god Thoth. This is, of course, part of the lore that surrounded Pythagoras. What is likely true is that Pythagoras derived his *musica universalis* theory from the Egyptians and the Chaldeans who held that the celestial bodies emitted a "cosmic chant" as they moved through the sky (Burkert 1972). Even in the Bible (Job 38.7), the beginning of time is described as the moment "when the stars of the morning sang together and all the sons of God raised a joyous sound."

The theory of *musica universalis* has always been an attractive one, as Jamie James (1995) has argued, because it portrays the universe as a stately, ordered mechanism—a "clockwork universe," as the cosmos has been described. Discoveries in biological science indicate that the *musica universalis* model may even describe organic systems. As biologists Bokai Zhu, Clifford C. Dasco, and Bert W. O'Malley (2018: 727) point out, musical rhythms are found not only in inanimate matter, but in biological systems as well:

> Contrary to traditional belief that a biological system is either at steadstate or cycles with a single frequency, it is now appreciated that most biological systems have no homeostatic "set point," but rather oscillate as composite rhythms consisting of superimposed oscillations. These oscillations often cycle at different harmonics of the circadian rhythm, and among these, the ~12-hour oscillation is most prevalent ... We posit that biological rhythms are also musica universalis: whereas the circadian rhythm is synchronized to the 24-hour light/dark cycle coinciding with the Earth's rotation, the mammalian 12-hour clock may have evolved from the circatidal clock, which is entrained by the 12-hour tidal cues orchestrated by the moon.

The key mathematical notion in *musica universalis* is that of commensurable *ratios*. What the Pythagoreans did not anticipate, however, was their unwitting discovery of numbers that were incommensurable, subsequently called *irrational*. So disturbed were they by this that, according to legend, they "went to the length of killing one of their own colleagues for having committed the sin of letting the nasty information reach an outsider" (Ogilvie 1956: 15). The colleague is suspected to have been Hipassus of Metapontum (Bunt, Jones, and Bedient 1976: 86). While this is likely to be a myth, the fact remains that the discovery of irrational numbers was mind-boggling, going (purportedly) against the *musica universalis* view of the cosmos.

Pi (π), or the ratio of the circumference of a circle to the diameter, was one of the first irrational numbers discovered. As such, it constituted a conundrum with regard to *musica universalis*. Pi represents a stable pattern in physical structure—it stands for the inescapable fact that, as the circumference of any circle increases, so too does its diameter, in a proportional way. As such, π would seem to fit in perfectly with the Pythagorean view of harmony, revealing an underlying pattern to an aspect of physical reality based on proportion. But the fact that π turns out to be an irrational number poses an existential paradox to the Pythagorean model of the world: Why is an incommensurable number connected to one of the most harmonious of all geometric forms, the circle? The mystery deepens by virtue of the fact that π has cropped up in various mathematical formulas that describe biological and physical structures and processes which, on the surface, seem to have nothing to do with circles.

So much has been written about π that it would be presumptuous to suggest that the themes I will be discussing in this book are novel (for comprehensive treatments see Beckmann 1971, Blatner 1997, Eymard and Lafon 2004, Posamentier 2004). Rather, my goal is to revisit the geometry of the Pythagoreans and its foundation as a "hermeneutic science," that is, a method of inquiry aiming to investigate the connectivity between geometric figures, numbers, and reality. The term *hermeneutics* (from Greek ἑρμηνεύω, *hermēneuō*) was introduced into philosophy by Aristotle in his book *Peri Hermeneias*, translated into Latin as *De Interpretatione*, and later in English as *On Interpretation* (Aristotle 2016). There is some suggestion that it may have a pre-Greek origin (Beekes 2009: 462). Whatever the case, it was Aristotle who made the cogent argument for the need to develop a philosophical method of "interpretation"—that is, a method for explaining the meanings of things. The geometry of the Pythagoreans can be retroactively described as hermeneutic, since they believed it would provide a key to decoding the meaning of the universe.

In the medieval period, the term *hermeneutics* designated specifically the interpretation of sacred scripture (Grondin 1994: 21); it continues to be used with this designation within theology to this day. Starting in the nineteenth century, the same term surfaced in philosophy as a theory of understanding, as can be seen in the writings of Friedrich Schleiermacher, Wilhelm Dilthey, Martin Heidegger, Hans-Georg Gadamer, among others (Seebohm 2007, Zimmerman 2015). It was not until the twenty-first century, however, that the term migrated to the domain of science and mathematics, where it is now used to identify any approach that connects the study of mathematics with humanistic disciplines, such as music and the arts. Although they did not identify it as such, George Lakoff and Rafael Núñez put forth one of the first comprehensive hermeneutic treatments of mathematics in their 2000 book, *Where Mathematics Comes From: How the Embodied Mind Brings Mathematics into Being* (2000). Lakoff and Núñez explained the ontological basis of mathematics as reflecting the same cognitive structures that underlie language, and especially metaphorical language. Since then, the hermeneutic approach in mathematics has spread broadly, especially in the educational sphere where it has had significant applications (see, for instance, Danesi 2019).

I must warn readers that they will not find in this book an in-depth or sophisticated treatment of the mathematics of π. In fact, I have made no technical-knowledge assumptions whatsoever. My goal is to revisit the Pythagorean idea that geometry may well be the code to unraveling the mystery (or mysteries) of existence. This book is actually inspired by Darren Aronofsky's brilliant movie, π: *Faith in Chaos* (1998), which is a cinematic essay on what hermeneutic geometry is essentially about—the contemplation of mathematical notions, such as π, as windows into the mysteries of existence. As the movie elucidates, π seems to crop up everywhere, in nature, human affairs, and physical matter, yet it remains mysterious. Mathematician W. W. Rouse Ball (1905: 23) recounts a famous anecdote related to the nineteenth century mathematician Augustus De Morgan that encapsulates the type of question that Aronofsky asks in the movie and which will be pursued in this book:

> De Morgan was explaining to an actuary what was the chance that a certain proportion of some group of people would at the end of a given time be alive; and quoted the actuarial formula, involving π [pi], which, in answer to a question, he explained stood for the ratio of the circumference of a circle to its diameter. His acquaintance, who had so far listened to the explanation with interest, interrupted him and exclaimed, "My dear friend, that must be a delusion, what can a circle have to do with the number of people alive at a given time?"

Marcel Danesi
University of Toronto, 2020

CHAPTER 1

Discovery of π and Its Manifestations

> The only real voyage of discovery consists not in seeking new land-
> scapes but in having new eyes.
> MARCEL PROUST (1871–1922)

∴

1 Prologue

Pi (π) is defined as the ratio of the circumference, C, of a circle (the distance around the circle) to its diameter, D (the linear distance across the circle passing through its center, O); equivalently, it is the ratio of the circumference to twice the radius (R):

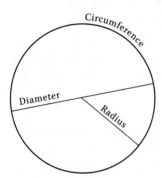

Definition: π = C/D or π = C/2R

In decimal form, the value of π is 3.14159265 …, with the digits after the decimal point continuing ad infinitum with, seemingly, no repeating pattern within them, unlike the *rational* numbers, whose digits show a recurring or finite pattern after the point:

1/2 = 0.5
1/4 = 0.25
1/3 = 0.3333 …
3/7 = 0.428571428571 … (in which 428571 repeats ad infinitum)

© KONINKLIJKE BRILL NV, LEIDEN, 2021 | DOI:10.1163/9789004433397_002

For this reason, π is defined as an *irrational* number—a number with an infinite, nonrecurring expansion after the decimal point. It is one of the first such numbers discovered in mathematical history. Its importance has been recognized since antiquity. As Beckmann (1971) observes, "the Babylonians and the Egyptians (at least) were aware of the existence and significance of the constant π," and both had numerical approximations of its value.

The Greeks used the letter π not for the ratio itself, but as a single-letter abbreviation for the word "periphery" (περιφέρεια, *periphéreia*), which also meant "circumference." This use of the symbol π continued right up to the seventeenth century. English mathematician William Oughtred, for instance, employed it in this way in his 1631 book *Clavis mathematicae*. However, as historian of mathematics Florian Cajori (1916), indicates, in a certain part of the book, Oughtred seems to have extended the meaning of the symbol to encompass its usage as a ratio:

> It is of some interest that Oughtred used π/δ to signify the ratio of the circumference to the diameter of a circle. Very probably this notation is the forerunner of the $\pi = 3.14159$ used in 1706 by William Jones.

The first attested use of π to represent the C/D ratio is in fact traced to William Jones's 1706 book, *Synopsis Palmariorum Matheseos, or a New Introduction to Mathematics,* as Cajori notes. In it, Jones indicates that he was influenced by British astronomer John Machin, who computed π to 100 decimal places—an amazing feat for the pre-computer era. The relevant section is the following one (Jones 1706: 243):

> There are various other ways of finding the lengths or areas of particular curve lines, or planes, which may very much facilitate the practice; as for instance, in the circle, the diameter is to the circumference as 1 to (16/ 5 − 4/239) − 1/3(16/53 − 4/2393) & c. = 3.14159 & c. This series (among others for the same purpose, and drawn from the same principle) I received from the excellent analyst, and my much esteemed friend Mr. John Machin.

The Machin-Jones notation was not adopted broadly by mathematicians at first; its widespread acceptance came after Leonhard Euler employed it in his 1736 work, *Mechanica*, and later in his *Introductio in analysin infinitorum* (1748). Given Euler's reputation as a leading mathematician of his era, this practice spread throughout European mathematics, with π becoming the standard symbol for the C/D ratio in all of mathematics subsequently.

It was shortly after this event that mathematicians and scientists started noting that π cropped up unexpectedly in formulas describing various physical phenomena that would seem to have nothing to do with the circle. As Michael Schneider (1994) has argued, it is one of those serendipitous symbols that surfaces by happenstance in nature and which seems to be part of an inherent system of principles that connects mathematics to reality. The study of this system is a *hermeneutic* enterprise, as discussed in the preface and as will be argued throughout this book. In the early medieval era the term was applied to the exegesis of sacred texts—an approach that is traced to Saint Augustine of Hippo (see 1887), who did not, however, explicitly use the word. In modern times, the term has spread more broadly to encompass the analysis of literary and cultural texts in terms of their specific linguistic features and historical cultural importance. In his book *Objective Knowledge* (1972), British philosopher Karl Popper claimed, for the first time, that the same hermeneutic principles and interpretive techniques could be applied as well to the study of scientific texts.

This opening chapter will discuss a few relevant historical facts about π. Although these are well known, it is nonetheless useful to revisit them here, so as to establish a frame of reference for discussing the scientific, artistic, and cultural meanings and uses of this ratio subsequently. It will also provide a generic characterization of "hermeneutic geometry," as a discipline whose goal is to apply geometric notions and models to investigate the connections that may exist among mathematical objects (figures, numbers, shapes, equations, etc.) and the elements of reality (natural and human).

2   Discovery, Calculation, Proof

The discovery of the ratio that π represents came about, in all likelihood, when the ancients began taking precise measurements of objects for practical reasons. Geometry was initially a craft for aiding construction, surveying, and other activities that required accurate measurement. The word *geometry*, which derives from the Ancient Greek *gē* "earth" and *metrein* "to measure," encapsulates what the ancient geometers did—they derived formulas from measuring lengths, land areas, and other dimensions that were part of everyday life. The Egyptians and Babylonians developed principles of elementary geometry in this way, laying the basis for "theoretical" geometry as a discipline concerned with the mathematical properties and interrelations that exist between geometric shapes, figures, and numbers. One of the earliest discoveries from early practical geometry was π.

Pi is mentioned twice in the Old Testament. In 2 Chronicles 4:2 we read: "Also he made a molten sea of ten cubits from brim to brim, round in compass and five cubits the height thereof, and a line of thirty cubits did compass it about;" and in 1 Kings 7:23, where we find an almost identical statement: "And he made a molten sea, ten cubits from the one brim to the other: it was round all about, and his height was five cubits: and a line of thirty cubits did compass it round about." These two citations tell us that the ancient Hebrews estimated π to be approximately 3—a cubit was a variable measure, based on the length of a human forearm (Engelson 2017). The Egyptians estimated it to be 3.1604. A Babylonian tablet from around 1900–1680 BCE gives the value of π as 3.125.

Calculating π to larger and larger values is no easy matter without a computer, because its digits after the decimal point have no finite or recurring pattern to them, as mentioned: 3.141592653589793 .... The quest to achieve increasingly accurate and lengthy calculations of π has been relentless—perhaps because of an unconscious hope that at some point the random digits will come to a halt or else some hidden pattern will finally be revealed the further along we compute its digits. The effort has not been a waste of time, though. In their calculations of π, mathematicians have stumbled upon a myriad hidden aspects of numbers and number systems that might not have been discovered otherwise. For example, in the fifteenth century, Indian mathematician Madhava of Sangamagrama devised an important infinite series for calculating π, which shows up in later mathematics as well. Although almost all of Madhava's original works are lost, his writings and ideas are discussed by later mathematicians of the Kerala School of Astronomy and Mathematics (which he had actually founded).

Madhava's formula for π was based on adding and alternately subtracting specific odd number fractions to infinity. German mathematician Gottfried Wilhelm Leibniz came up with the same formula two centuries later, whence it came to be known as the Leibniz-Madhava Series (Gupta 1975: 46)—as far as can be told this is a case of serendipitous coincidence, since it is improbable that Leibniz would have had access to the original Kerala works that cite Madhava:

Madhava Series:
$\pi = 4 - 4/3 + 4/5 - 4/7 + 4/9 - 4/11 + 4/13 - 4/15 \ldots$
Leibniz-Madhava Series:
$\pi/4 = 1 - 1/3 + 1/5 - 1/7 + 1/9 - 1/11 + 1/13 - 1/15 \ldots$
[Note that this is the same series divided by 4]

The above series is sometimes called the Madhava-Leibniz-Gregory Series, because it is a special case of the series for computing the inverse tangent function (known as arctan) discovered by the seventeenth-century Scottish mathematician James Gregory:

$$\text{Arctan } x = x - x^3/3 + x^5/5 - x^7/7 + x^9/9 \dots$$

Note that the Leibniz-Madhava Series results when $x = 1$ in the Gregory series. Adding up the terms in any of these series allows us to compute the value of π to as many digits as we desire by simply moving further and further up the series. Today, computers can calculate π to trillions of digits, allowing mathematicians to examine not only the digits in themselves, but also, and more significantly, the properties of the algorithms employed to generate them as potentially revealing a hidden pattern. So far, the search for a pattern in π has been to no avail.

In an episode of the original *Star Trek* series on American television, called "The Wolf in the Fold" (1967), the character Mr. Spock used the randomness of the digits of π as a ploy to defeat a computer gone amok. He simply commanded the computer to find the final digit of π, knowing full well that the computer would not stop churning out digits and thus would eventually break down. Like the computer, it seems that we humans also cannot leave π alone, trying to find a way somehow to arrive at the final digit, or else venture off into infinity along with the digits—perhaps even to the point of losing our own minds, as the movie π: *Faith in Chaos*, mentioned in the preface and to be discussed in Chapter 5, suggests rather powerfully. Efforts to calculate π have not, however, been a waste of time and effort, as discussed above. These have led to discoveries that would have otherwise remained unknown, such as the Madhava-Leibniz-Gregory Series. A world without π is, of course, conceivable. But what we now know about the sun and the tides, among many other natural phenomena, would be much more rudimentary. As Kasner and Newman (1940: 89) have aptly put it, without π "our ability to describe all natural phenomena, physical, biological, chemical or statistical, would be reduced to primitive dimensions."

One of the first formal proofs of the value of π is found in a work dating back to 1650 BCE called the *Ahmes Papyrus*, after the Egyptian scribe Ahmes who copied it from an earlier work. It is also called the *Rhind Papyrus*, after the Scottish lawyer and antiquarian, A. Henry Rhind, who purchased it in 1858 while traveling in Egypt. The papyrus had been found a few years earlier in the ruins of a small building in Thebes in Upper Egypt. On the basis of what

Ahmes writes in the preface, the original work can be traced back to the 1800s BCE (Chace 1979: 27):

> This book was copied in the year 33, in the fourth month of the inundation season, under the majesty of the king of Upper and Lower Egypt, 'A-user-Re', endowed with life, in likeness to writings made in the time of the king of Upper and Lower Egypt, 'Ne-ma'et-Re'. It is the scribe A'h-mose who copies this writing.

A-user-Re has been identified as a member of the Hyskos dynasty, which governed around 1650 BCE—the year that Ahmes was copying the work. Ne-ma'et-Re was Amenemhet III, who reigned from circa 1860 to 1814 BCE, the time frame during which the original work was written. In addition to over eighty mathematical problems, the papyrus contains tables for the calculation of areas, the conversion of fractions, elementary sequences, linear equations, and detailed information about measurement. The earliest known symbols for addition, subtraction, and equality are also found in this extraordinary work (Gillings 1961, 1962, 1972). It was translated into German in 1877 by mathematician August Eisenlohr, and into English in 1923 by British scholar Thomas Eric Peet. The first extensive annotated edition of the work was completed in 1929 by Arnold Buffum Chace (see Chace 1979)—an edition that made the papyrus accessible for the first time to a broad audience. The original papyrus is now preserved in the permanent collection of the British Museum, which came into its possession after Rhind's death. It was missing certain fragments, but by a stroke of good fortune, these were found a little later in the guardianship of the New-York Historical Society.

The work starts off with the following enigmatic sentiment:

> Accurate reckoning, the entrance into the knowledge of all existing things and all obscure secrets.

The phrase "obscure secrets" is a metaphor for unknown ideas that can be discovered with "accurate reckoning," a plausible characterization of how mathematics unfolds. The work was probably a school textbook, aiming to get students to think mathematically and learn about the practical things that mathematics allowed them to do (Gillings 1961, Olivastro 1993: 31–64). But an argument can be made that it was an early treatise in mathematics as well. Its title—*Directions for Attaining Knowledge of All Dark Things*—alludes, presumably, to the belief that mathematics and mystery were intertwined. It could thus mean that mathematics provides the set of "directions" for decoding the "dark" (unknown) things of the world.

The work contains one of the first proofs of the value π in history, presented as the solution to a problem, which can be paraphrased as follows:

What is the area of a circle inscribed in a square that is 9 units on its side?

The author of the papyrus solved it with the following truly brilliant insight, which can be rendered as follows: What if an octagon is inscribed inside the square? It is easy to calculate the areas of the two polygons (square and octagon), which we can then utilize to approximate the circle's area. To inscribe an octagon inside the square, the author suggested trisecting each side of the square into nine smaller squares, (each 3 × 3), as shown in the diagram below, and then drawing diagonals in the corner squares. As can be seen, such modifications produce an octagon, whose area will approach that of the circle:

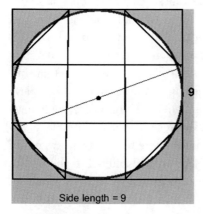

Side length = 9

Diagram for calculating π
SOURCE: WIKIMEDIA COMMONS

Now, the area of the octagon is equal to the areas of the five smaller inner squares (which form the outline of a cross) plus half the areas of the four corner squares (= the area of two squares). This means that the area of the octagon is equal to the sum of the areas of seven small squares in total. The area of one of these is 3 × 3, or 9 square units. The total area of seven such squares is, therefore, 9 × 7, or 63 square units. Now we can use this information, along with the fact that the diameter is 9 units, since it is equal in length to a side of the square, as can be seen in the diagram, in order to estimate the area of the circle and, as a derivative, the value of π as follows (using modern notation):

Diameter = 9
Radius (r): = 9/2

So, r^2 $= (9/2)^2 = 20.25$
Area of circle $(\pi r^2) = \pi \times 20.25$
Area of octagon $= 63$
Area of circle $=$ Area of octagon (approximately)
$\pi \times 20.25 = 63$
$\pi = 63/20.25$
Since the circle is a little larger than the octagon, we can assume it to be around 64 units; therefore:
$\pi = 64/20.25 = 3.1604 \dots$

This proof is truly extraordinary for the era in which it was devised, since it embodies an ingenious idea—the area of the circle and that of a polygon do not coincide perfectly, but for practical purposes they can be imagined to be close enough. This bears within it the intellectual seeds for the concept of limits, which subsequently became the root of several mathematical branches, including the calculus. A *limit* is defined technically as a point or value that a function, figure, or series can be made to approach progressively, until it is as close to the point or value as desired. What is especially noteworthy is that similar methods of proof to the one used in the papyrus surfaced in diverse parts of the ancient world. It is unlikely that the ancient mathematicians would have been aware of each other's proofs, since they were written in different languages and in disparate parts of the globe in an era where there was no system of mass communications. This suggests that the proof exhibited in the *Ahmes Papyrus* was based on an unconscious geometric archetype, namely "squaring-the-circle," which envisions a circle as a polygon, starting as a square that can be smoothed out by infinitely increasing the number of its sides to approach the curvature of a circle.

The modern-day meaning of the term *archetype* comes, as is well known, from Jungian psychology (Jung 1983). It refers to any pattern of thought that is embedded in the psyche manifesting itself across time and cultures in specific ways; it is the cognitive counterpart of instinct. So, in a sense, "squaring-the-circle" is an archetype that surfaces wherever geometry is practiced. The association between circles and squares probably came from a common experience throughout the ancient world of measuring these two shapes and making analogies between them. So widespread was this archetype that it even produced a fallacy, namely the belief that a square could be constructed with the exact same area of a given circle by using only a finite number of steps with compass and straightedge. This was proven to be impossible in 1882, when it was demonstrated that π is a transcendental, rather than algebraic, irrational number; that is, it is not the root of any polynomial with rational co-efficients (to be discussed subsequently). This proof is called the Lindemann–Weierstrass theorem (Baker 1990).

The technique of squaring the circle to any given non-perfect degree of accuracy is possible, as the Ahmes proof showed, in a finite number of steps, since there are rational numbers arbitrarily close to π. It is a kind of proof that involves analogical thinking, which itself has produced many other discoveries in mathematics, as Hofstadter and Sander (2013) have cogently argued. This type of analogical proof also reveals what Charles S. Peirce, a mathematician and logician, called *abduction*. He described it as follows (Peirce 1931–1958, V: 180):

> The abductive suggestion comes to us like a flash. It is an act of *insight*, although of extremely fallible insight. It is true that the different elements of the hypothesis were in our minds before; but it is the idea of putting together what we had never before dreamed of putting together which flashes the new suggestion before our contemplation.

The relevant "abduction" in the Ahmes proof is that the area of a circle can be envisioned as the smoothed-out area of an octagon inscribed in a circle bounded by a square. Anecdotal corroboration that the proof is archetypal is the fact that the same abduction occurred to mathematicians in other parts of the world. The most widely-cited one is by Archimedes (Arndt and Haenel 2006). His method can be encapsulated as follows. Consider a circle with diameter equal to 1 (radius = ½). First, inscribe a regular hexagon in it, in such a way that its six vertices bisect the arcs of the circle:

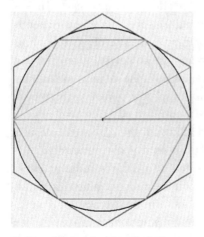

Archimedean method for calculating π

A regular hexagon outside the circle can now be drawn, as shown in the diagram. We now have an inscribed and a circumscribed hexagon. The perimeter of each can be easily calculated, with the perimeter of the inscribed hexagon smaller than the circumference of the circle and the perimeter of the circumscribed hexagon larger than the circumference. Now, by increasing the number of sides of successive inscribed and circumscribed polygons, the perimeters will converge towards the value of the circumference. Archimedes started with inscribed and circumscribed hexagons (as above), then doubled the number of sides of each to produce 12-sided regular polygons, inside and outside the circle. Using the same iterative method, he inscribed and circumscribed 24, 48, and 96-sided polygons (see Heath 1958).

In this way, Archimedes was able to bring the perimeters of the inscribed and circumscribed polygons closer and closer to the circumference of the circle, leading to his approximation of π as lying between 3 and 10/71 and 3 and 1/7. Similar iterative methods have been discovered across societies and eras, suggesting that the iterative method is itself an offshoot of the squaring-the-circle archetype. For example, the Chinese mathematician Lin Hui (c. 250 CE) used a variation of Archimedes' method to determine π as equal to 3.14159 (Kangshan, Crossley, and Lun 1999). Around 400 CE, in an Indian manuscript, referred to as the *Surya Siddhanta* ("System of Astronomy"), the value for π is given as 3.1416, again using the same type of iterative method (Plofker 2009).

3   Geometric Archetypes

The notion of *geometric archetype* requires justification, since it is a central one in elaborating a theory of geometry as a hermeneutic science—hermeneutic geometry for short. It can be defined as any insight, figure, or method of proof that transcends time and place. So, the iterative proofs for π are archetypal, given that they are found across cultures. Moreover, they are based on the same abduction that envisions a resemblance between polygons and circles. In semiotics, this type of thinking is called *iconic*, starting with Peirce (mentioned above), who characterized iconicity as the tendency to understand unknown and unnamed objects in terms of how they resemble some perceived property in analogous objects.

So, connecting a polygon (with infinite sides) to a circle is, more precisely, an iconic geometric archetype that is based on imagining similarities in geometric shapes through analogical thinking. The term *transformation* is a central one in mathematics, defined as the process by which one figure or function is converted into another that is equivalent in some respect but is differently represented. The transformation of a polygon into a circle can plausibly be explained, therefore, as an iconic geometric abduction that involves a universal archetype. This is perhaps the reason that, when we are presented with a proof such as the one by Ahmes, we require no special logical justification for it. We respond to it archetypally, not skeptically. This meaning of archetype can be enlisted more generally to explain the discovery of recurrent patterns in mathematics and their manifestations in the real physical world, as will be discussed throughout this book (Schneider 1994, Adam 2004, Danesi 2018).

The notion of geometric archetype is not limited to analogies or similarities perceived among figures and shapes but, in some cases, to the shapes themselves. In this sense, the circle is a geometric archetype, given that is found throughout the world and across time. So too is the square. Thus, the term *geometric archetype* has two designations: (1) it refers to specific shapes, such as circles and squares, that are found universally; and (2) it implies imagining relationships among geometric shapes, so that one can be turned into another, as in the squaring-the-circle abduction. In the case of the circle, the type-(1) archetype above is, more precisely, *circularity* or *curvature*; and in the case of the square it is *edge-ness*. Type-(2) archetypal thinking is the method of envisioning something as connected to something else. So, the above iterative proofs of π are based on the type-(2) archetypal transformation of edge-ness into circularity.

Archetypes permeate geometry. Consider the so-called golden ratio, which results from the division of a line made up of two segments, a and b, in such a way that the ratio of the whole line, $(a + b)$, to the larger part, a, is equal to the ratio of that part, a, to the smaller part, b—that is, $(a + b)$ is to a as a is to b. The ratio, represented with the Greek letter *phi* (ϕ), is approximately 1.61803. The line below is divided into this proportion:

The golden ratio
SOURCE: WIKIMEDIA COMMONS

Other archetypal geometric figures can be designed with this ratio. For example, a golden rectangle with sides $(a + b)$ and a has this ratio:

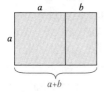

The golden rectangle
SOURCE: WIKIMEDIA COMMONS

Historians trace the formal study of the golden ratio to the Greek sculptor and mathematician Phidias (500–432 BCE), who applied it to the design of sculptures for the Parthenon, although it is recorded that the Egyptians also knew about it, calling it the "sacred ratio." The Pythagoreans were also aware of ɸ and, according to one legend, Hippasus (a member of the cult) discovered that it was an irrational number, surprising the members as much as their unwitting discovery that their right triangle theorem also produced irrational numbers (Livio 2003). Euclid's *Elements* (see 1956) contains several proofs employing the golden ratio and the first known definition:

> A straight line is said to have been cut in extreme and mean ratio when, as the whole line is to the greater segment, so is the greater to the lesser.

The golden ratio was subsequently used by Egyptian mathematician Abu Kamil (c. 850–930) to study polygons; Italian mathematician Leonardo Fibonacci (c. 1170–1250) also employed the ratio in formulating several geometry ideas which may, in turn, have influenced Luca Pacioli (c. 1447–1517) to study the sense of aesthetic beauty that results when this ratio is employed in architecture and art in his book *Divina proportione* (1509). Leonardo da Vinci illustrated Pacioli's book, calling ɸ the *sectio aurea* (golden section). Many hidden patterns associated with ɸ have since been documented, as well as many serendipitous appearances of this ratio in nature and human affairs (as will be discussed subsequently).

4   Manifestations of Archetypal Structure

In the 1600s, π surfaced in the formulas of a whole series of curves, suggesting an archetypal affinity between circularity and curvature, as mentioned above. Pi has also been found to occur in number theory, probability theory, chaos theory, topology, trigonometry, the calculus, and on and on, suggesting the presence of a curvature archetype in many domains of mathematics that the appearance of π fleshes out. Topology concerns itself with determining properties of shapes and configurations, such as *insideness* or *outsideness*. A circle, for instance, divides a flat plane into two regions, an inside and an outside. A point outside the circle cannot be connected to a point inside it by a continuous path in the plane without crossing the circle's circumference. If the plane is deformed, it may no longer be flat or smooth, and the circle may become a crinkly curve, but it will continue to divide the surface into an inside and an outside. Circularity can thus be seen as a specific type of curvature.

As mentioned, type-(2) archetypal thinking dovetails with that of *iconicity* in semiotics (Sebeok and Danesi 2000). Imagining a multi-sided polygon as "evolving" into a smooth curve via iterative transformations is a product of iconic thinking. The Estonian biologist, Jakob von Uexküll (1909), called this type of cognition an innate "modeling system," whereby humans envision certain properties in the world as connected via perceived resemblances, transforming them into models that concretize the connectivity.

Now, it can be argued that the mathematical notion of infinity can also be classified under the rubric of a type-(1) geometric archetype. It involves envisioning something, such as a straight line, as having no end—an image that is found across cultures and eras. This archetype shows up concretely in a series of famous ancient paradoxes, formulated by Zeno of Elea (Salmon 1970, Mazur 2008). With them, Zeno aimed to demonstrate that motion, change, and plurality (reality consisting of many substances) are theoretically impossible, in support of the doctrine of his teacher, Parmenides. His Achilles and the Tortoise Paradox is perhaps the best known one. It is recounted in Aristotle's *Physics* (see Aristotle 2015). Below is a paraphrase:

> Achilles decides to race against a tortoise. To make the race fairer, he allows the tortoise to start at half the distance away from the finish line. In this way, Achilles will never surpass the tortoise. Why is this so?

The argument utilized is an exemplification of *reductio ad absurdum* reasoning ("reduction to the absurd") (Rescher 2009). In order for Achilles to surpass the tortoise, he must first reach the halfway point, which is the tortoise's starting point. But when he does, the tortoise will have moved forward a bit. Achilles must then reach this new point before attempting to surpass the tortoise. When he does, however, the tortoise has again moved a little bit forward, which Achilles must also reach again, and so on ad infinitum. In other words, although the distances between Achilles and the tortoise will continue to get smaller, infinitesimally so, Achilles will in theory never surpass the tortoise. Of course, in reality Achilles will do so, as a real-world race between any human and a tortoise would certainly confirm.

Another of Zeno's significant paradoxes, called the Dichotomy Paradox, states that Achilles will never reach the finish line because, before he can get there, he has to reach the halfway point first. Then, from there he must go another half of the remaining distance (or another quarter of the way) to the end. From there, he has to travel another half distance (or one-eighth of the way). Subsequently, he must cover distances of one-sixteenth, one-thirty-second, and so on ad infinitum. So, Achilles will not never get to the finish line, approaching it ever so closely, but not quite. The different phases of Achilles' motion form an infinite series with each term in it half of the previous one:

$$\{1/2 + 1/4 + 1/8 + 1/16 + 1/32 + ...\}$$

Like the iterative method used for proving the value of π (above), Zeno's paradoxes imply the concept of limits. The calculus subsequently solved the paradoxes ingeniously, by proposing the idea of *infinitesimals*, which are entities that are so small that there is no way to measure them, although they retain specific properties, such as angle or slope. The concept was introduced around 1670 by Leibniz (Katz and Sherry 2012). However, as discussed above, it was implicit in the iterative method used by Archimedes and others, who did not specifically measure the sides of increasingly-large polygons circumscribed around and inscribed in a circle to estimate the value of π, but rather gave infinitesimal approximations of them. Archimedes used the same technique in his book, *The Method of Mechanical Theorems*, to find areas of regions and volumes of solids (Heath 1958, Askew and Ebutt 2011). It was John Wallis who introduced the symbol ∞ to stand for infinity in his *Treatise on the Conic Sections* (1655), enshrining the archetype of infinity into mathematics as a distinct notion (Scott 1981).

Although they do not name them as such, George Lakoff and Rafael Núñez (2000) have cogently argued that abductions such as those discussed above result from blending different concepts and connecting them via verisimilitude—a process that is behind metaphor in language. For example, some mathematical ideas might result from what Lakoff and Núñez call conceptual metaphors, which arise by blending specific domains of experience to mathematics. So, addition develops from the experience of counting objects and imagining them as being inserted in a collection, such as a box or some other container, or else holding them in the hands. This is now called a *conceptual* (or *cognitive*) *blend* (Fauconnier and Turner 2002). Blends are the purported sources of many mathematical notions. One of these is negative numbers, as Alexander (2012: 28) elaborates:

> Using the natural numbers, we made a much bigger set, way too big in fact. So we judiciously collapsed the bigger set down. In this way, we collapse down to our original set of natural numbers, but we also picked up a whole new set of numbers, which we call the negative numbers, along with arithmetic operations, addition, multiplication, subtraction. And there is our payoff. With negative numbers, subtraction is always possible. This is but one example, but in it we can see a larger, and quite important, issue of cognition. The larger set of numbers, positive and negative,

is a cognitive blend in mathematics ... The numbers, now enlarged to include negative numbers, become an entity with its own identity. The collapse in notation reflects this. One quickly abandons the (minuend, subtrahend) formulation, so that rather than (6, 8) one uses -2. This is an essential feature of a cognitive blend; something new has emerged.

Blending, as Alexander suggests, is the cognitive source for expanding mathematical paradigms. Once π was discovered as a ratio connected with the circle form, it was applied to the study of other geometrical objects that possessed the archetype of curvature. So, for example, armed with π we can now extend it to determining the volume of a cylinder, in which the area of the base circle is obtained in the usual fashion by multiplying the square of the radius by π; that is, πr^2:

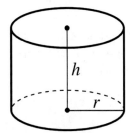

Volume of a cylinder

The height of the cylinder is h, so the volume is $\pi r^2 h$. Now, this formula, allows us to use it to calculate the volume of any cylinder easily. The volumes or areas of other geometric figures that are based on the curvature archetype are the following:
- Volume of a cone: $1/3\ \pi r^2 h$
- Volume of a sphere: $4/3\ \pi r^3$
- Lateral area of a cylinder: $2\pi rh$
- Surface area of a cylinder: $2\pi rh + 2\ \pi r^2$
- Surface area of a sphere: $4\pi r^2$

The shift from practical to theoretical geometry, from the specific to the general, is the core of mathematics, and it is guided initially by archetypal reasoning. This is why the same kind of mathematics shows up in different guises in other parts of the world at other times, transcending geographical and cultural spaces. In other words, archetypes such as the curvature (or circularity), edge-ness, and infinity ones produce patterns of thought that transcend culture-specific reasoning.

Before conceptual blending theory, philosopher Max Black (1962) claimed that metaphors revealed how scientists and mathematicians think archetypally, since they produce mental images of phenomena that we cannot see with our eyes—atoms, sound waves, gravitational forces, magnetic fields, and so on. As scientist Robert Jones (1982: 4) observed, for the scientist metaphors serve as "an evocation of the inner connection among things." In a similar vein, Fernand Hallyn (1990) argued that scientific models are poetic in the sense that they allow us, like poetry, to visualize things as interconnected. It is by using this kind of metaphorical (connective) inner vision that we come to contemplate principles of reality beyond instinct and intuition, such as the presence of circularity and iteration in various domains of reality.

Consider Buffon's Needle Problem, first posed in the eighteenth century by Georges-Louis Leclerc, Comte de Buffon (1777: 100–104). The problem can be paraphrased as follows:

> Suppose there is a floor on which parallel lines are drawn, all the same width apart. A needle equal in length to the width between the lines is dropped randomly onto the floor a large number of times. The needle will either fall between parallel lines or else cross one of them. What is the probability that the needle will cross one of the lines?

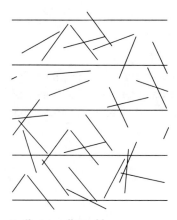

Buffon's needle problem
SOURCE: WIKIMEDIA COMMONS

It turns out that, as the number of tosses of the needle increases, the ratio of the number of tosses to the number of times the needle crosses a line approaches π. Margaret Willerding (1967: 120) recounts a historical anecdote whereby in 1901 "a scientist made 3408 tosses of the needle and claimed that it touched the lines 1085 times," and thus the "ratio of the number of tosses to the number of times the needle touched the lines, 3408/1085, differs from π by less than 0.1 per cent." This is nothing short of perplexing given that one can ask: Where is circularity in this experiment? Buffon's Needle Problem has been a topic of debate and discussion within mathematics since its formulation (for example, Schroeder 1974). However, if we represent the problem graphically, what emerges is a probability curve, suggesting indirectly that the curvature archetype is embedded deeply within the problem (for example, Kulyukin 2007). Without going into the mathematical details here, suffice it to say that when the results of Buffon's Needle experiment are plotted on a coordinate system, curves appear, suggesting that it is through graphical representation that the curvature archetype is fleshed out and made obvious as being somehow embedded in the experiment. If one stops to think about it, this is a truly remarkable finding—namely, the representational and notational techniques of mathematics lead to discoveries in themselves. It is a virtual law of mathematical discovery that any new notational device, symbol, or technique will stimulate further ideas and suggest new paths of discovery.

5   Geometry as a Hermeneutic Science

The Pythagoreans were among the first to envision geometry as a hermeneutic tool for exploring the connection between geometric ideas and aspects of reality—a perspective of geometry adopted subsequently by Aristotle and Plato (Dancy 2007, Pettigrew 2009). The term *hermeneutic geometry* is intended to encapsulate the Pythagorean view, providing a semiotic-interpretive framework for grasping the connectivity itself—hence a "method for interpreting connections." In this framework, a notion such as circularity comes not only from seeing circles of various dimensions, but from imagining certain shapes as possessing a similar structure, even if deformed topologically in some way (as discussed above). Similarly, the notion of infinity comes from imagining how lines, for instance, can be extended without end.

Now, a further extension of hermeneutic geometry would be to examine connections among the specific archetypes themselves in order to see if there is an overarching connective structure involved. A linkage between the archetypes of circularity and infinity was established by the Madhava-Leibniz-Gregory Series—that is, as a formula for calculating π (a ratio intrinsic to the circle) it shows that the ratio can only be approximated in terms of an infinite sequence. Here are two other famous infinite series, in addition to the Madhava-Leibniz-Gregory one, that have been used to calculate π as a limit:

$$\pi/2 = (2/1 \times 2/3) \times (4/3 \times 4/5) \times (6/5 \times 6/7) \dots$$
$$\pi^2/6 = 1/1^2 + 1/2^2 + 1/3^2 + 1/4^2 + 1/5^2 \dots$$

The top one was devised by John Wallis (1656); it shows that π can be calculated in terms of an infinite series consisting of the product of consecutive terms with the following structure:

$\pi/2 = (2n \,/\, 2n - 1) \times (2n \,/\, 2n + 1)$, where n goes from 1 to infinity

The symbol $2n$ stands for any even number, while both $2n - 1$ and $2n + 1$ stand alternatively for any odd number:

First term $(n = 1)$: $(2n \,/\, 2n - 1) \times (2n \,/\, 2n + 1) = (2/1 \times 2/3)$
Second term $(n = 2)$: $(2n \,/\, 2n - 1) \times (2n \,/\, 2n + 1) = (4/3 \times 4/5)$
Third term $(n = 3)$: $(2n \,/\, 2n - 1) \times (2n \,/\, 2n + 1) = (6/5 \times 6/7)$
...

Adding up the terms in the series as n becomes larger and larger, the series approaches the value of π.

It is intriguing to note how the most fundamental of all classification in mathematics—even versus odd numbers—is also somehow connected to the π ratio. The Wallis Product, as it came to be called, has had implications for trigonometry, the calculus, and other domains of mathematics and science, which need not concern us at this point.

The second series above was discussed by Leonhard Euler in 1764. It consists of terms with a simple structure, namely $1/n^2$, where n is the number of the term in the series:

First term $(n = 1)$: $1/n^2 = 1/1^2$
Second term $(n = 2)$: $1/n^2 = 1/2^2$
Third term $(n = 3)$: $1/n^2 = 1/3^2$
...

Again, adding up the terms in the series as n becomes larger and larger, the series approaches the value of π.

What is truly remarkable is that these series somehow produce values of π. As Mackenzie (2012: 44) insightfully points out:

These equations reveal that the number pi is not merely a geometric concept. Three of the great tributaries of mathematics merge in these formulas: geometry (the number pi), arithmetic (the sequence of odd numbers, and the sequence of squares 1^2, 2^2, 3^2, ...), and analysis of the infinite (in this case, infinite sums). Archimedes would have been flabbergasted to see formulas like these.

As we saw above, π is also connected to probability functions, as evidenced by its appearance in the graph of Buffon's Needle Problem. Pi thus seems to be an element in several archetypal structures, including circularity, curvature, infinity, and probability, thus suggesting that these archetypes are somehow interconnected. The same line of reasoning applies to other mathematical discoveries, such as *e*. Ian Stewart (2008: 101) synthesizes this whole situation as follows:

> The number *e* is one of those strange special numbers that appear in mathematics, and have major significance. Another such number is π. These two numbers are the tip of an iceberg—there are many others. They are also arguably the most important of the special numbers because they crop up all over the mathematical landscape.

As mentioned at the start of this chapter, the term *geometry* initially described what the ancient engineers and builders did. They measured the size of fields, laid out accurate right angles for the corners of buildings, and carried out other such practical mensuration activities. They used diagrams to represent their calculations and their layouts. By the seventh century BCE, the Greek mathematician Thales of Miletus established proof in geometry as a means to extract general principles of geometric structure from such activities. Pythagoras adopted the method of proof to show, among other things, that there was a consistent relation among the sides of right-angled triangles. By 300 BCE, Euclid established geometry as a system of proofs derived from basic axioms. At that point, geometry had evolved into a discovery tool that could be used to investigate reality, deriving formulas and notions that made it possible to explore the world in the mind, without having to intervene physically.

A classic example of the latter is the measurement of the circumference of the earth by Greek geometer Eratosthenes (c. 275–195 BCE)—a truly mind-boggling achievement showing how powerful abstract geometric reasoning can be. Standing during the summer solstice (June 20) at Alexandria, and knowing that it was due north of the city of Syene, with the distance between the two cities being 5000 stadia (500 miles), Eratosthenes was able to calculate the earth's circumference, without having to do it physically. One version of the story, reduced considerably, goes somewhat as follows. Eratosthenes saw that the noon sun's rays directly above cast a shadow off a pole near him. He proceeded to measure the angle between the rays and the pole as 7.2°. The sun's rays also shined straight down into a well at the same time. Since the rays are parallel, by a theorem of Euclidean geometry, Eratosthenes knew that the angle between them was also 7.2°. Reasoning further, he knew that this angle was 7.2° of 360° (since the earth is virtually a sphere). He concluded that the circumference divided by the distance between the two cities (5,000 stadia) thus formed the proportion 360°/7.2°:

Circumference/5,000 stadia = 360°/7.2°
Therefore:
Circumference = 360°/7.2° × 5,000 = 250,000 stadia

This is in close agreement with the actual known value today of approximately 24,901 miles.

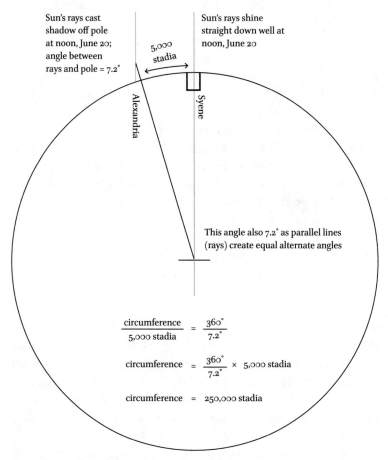

Eratosthenes' measurement of the earth's circumference
SOURCE: SIMON FRASER UNIVERSITY ONLINE

Ian Stewart (2008: 46) summarizes the prodigious achievements that geometry has permitted as follows:

Using geometry as a tool, the Greeks understood the size and shape of our planet, its relation to the Sun and the Moon, even the complex motions of the remainder of the solar system. They used geometry to dig long tunnels from both ends, meeting in the middle, which cut construction time in half. They built gigantic and powerful machines, based on simple principles like the law of the lever.

As discussed, it was Aristotle who first used the term *hermeneutics*. The etymology of the term is, however, not clear. A speculative version is that it is derived from the name of Hermes, the messenger of the gods and the inventor of language. Now, to claim that ancient geometry was a hermeneutic science requires a specific definition, especially since, as Rebecca Stephenson (2011: 108) aptly points out, the word "hermeneutic" today is used too broadly, watering it down to mean anything that involves interpretation. So, to avoid vagueness, the term *hermeneutic geometry* is defined as the systematic study of the connections between geometry, other fields of mathematics, nature, and human cultural artifacts, using notions such as archetype, abduction, and others. Circularity and infinity are examples of connective archetypal structures. In short, hermeneutic geometry examines the presence of geometric notions, such as π, and how these appear archetypally in all kinds of seemingly disparate phenomena and systems.

In the late nineteenth century, and into the early twentieth, the term hermeneutics was used to imply the development of an overall theory of understanding (Seebohm 2007, Heidegger 2008). Within this paradigm, Karl Popper (1972) devised the method of "objective hermeneutics," whose goal was to provide an interpretive framework for uniting all disciplines, as stated explicitly by Oevermann, Allert, Konau, and Krambeck (1987: 436–437):

Our approach has grown out of the empirical study of family interactions as well as reflection upon the procedures of interpretation employed in our research. For the time being we shall refer to it as objective hermeneutics in order to distinguish it clearly from traditional hermeneutic techniques and orientations. The general significance for sociological analysis of objective hermeneutics issues from the fact that, in the social sciences, interpretive methods constitute the fundamental procedures of measurement and of the generation of research data relevant to theory. From our perspective, the standard, nonhermeneutic methods of quantitative social research can only be justified because they permit a shortcut in generating data (and research "economy" comes about under specific conditions). Whereas the conventional methodological attitude in the social sciences justifies qualitative approaches as exploratory or preparatory activities, to be succeeded by standardized approaches and techniques as the actual scientific procedures (assuring precision, validity, and objectivity), we regard hermeneutic procedures as the basic method for gaining precise and valid knowledge in the social sciences. However, we do not simply reject alternative approaches dogmatically. They are in fact useful wherever the loss in precision and objectivity necessitated by the requirement of research economy can be condoned and tolerated in the light of prior hermeneutically elucidated research experiences.

In this book, the term hermeneutic geometry is used in line with objective hermeneutics; it is however constrained to the study of the geometrical meanings of phenomena, that is, the analysis of the connections between geometric archetypes and their manifestations in other areas. The many serendipitous appearances of π in science, for example, would be a target of hermeneutic geometry; so too would its manifestations in architecture and art. Hermeneutic geometry would thus aim to examine points of contact among mathematics, science, the arts, human history, etc. as elements in human understanding. It would thus necessarily involve a crisscrossing among several main disciplines—geometry, mathematics (generally), semiotics, psychology, philosophy, and anthropology. The study of abstract pattern is what mathematics is all about. Semiotics studies how such pattern is represented and how it is linked to cultural practices. Philosophy aims to explain a geometric form in broad intellectual terms. Psychology can be enlisted to shed light on what occurs in the mind as we reason geometrically. Anthropology helps understand the cultural-contextual factors that guide the development of geometric ideas.

A distinction between theoretical geometry and hermeneutic geometry is essential at this point. The former is geometry proper. It is a branch of mathematics that studies phenomena of shape, size, relative position, and properties of space. It arose independently in a number of early societies as a way for dealing with problems of engineering, as discussed several times. It became a formal theoretical discipline with Thales, Pythagoras, Euclid, and a few other ancient mathematicians. The amalgamation of geometry and algebra occurred in the work of René Descartes and Pierre de Fermat in the sixteenth and seventeenth centuries. Since then, geometry has been expanded into non-Euclidean branches, describing concepts that lie beyond the normal range of human experience. Hermeneutic geometry focuses on the connections between geometry and other creative activities, including science an art.

Recall the example of the golden ratio (above). Although known throughout antiquity, it was at the beginning of Book VI of the *Elements* that Euclid finally made it part of geometry proper. However, the ratio has hardly remained within the strictly Euclidean paradigm; even in antiquity it stimulated the interest of philosophers, artists, and architects, since it seems to have a subtle aesthetic effect on people when it is incorporated into art or architecture. The rectangular face of the front of the Parthenon in Athens has sides whose ratio is golden. The ratio of the height of the United Nations Building in New York (built in 1952) to the length of its base is also golden. Both buildings have been shown to have a particular beauty that is appreciated by a wide variety of people (Livio 2002).

Now, consider the golden rectangle, which can be shown to embed both the infinity and iterative archetypes, by using the so-called Fibonacci Sequence (Fibonacci 1202). The Fibonacci Sequence, as is well known, is an infinite series in which a term is the sum of the previous two:

$$\{1, 1, 2, 3, 5, 8, 13, 21, 34, 55, 89, 144, 233, 377, 610, 987, ...\}$$

So, for example, 2 (the third number) = 1 + 1 (the sum of the previous two); 3 (the fourth number) = 1 + 2 (the sum of the previous two); etc. This pattern can be generalized by using appropriate notation. If we let Fn represent any "Fibonacci number," and Fn_{-1} the number just before it, and Fn_{-2} the number just before that, the pattern inherent in the sequence can be shown as follows:

$$Fn = Fn_{-1} + Fn_{-2}$$

This is a recursion formula that provides a snapshot of the internal structure of the sequence. Recursion is a type of iteration, thus connecting the Fibonacci Sequence itself to archetypal structure. Now, the remarkable thing is that in the golden rectangle, we can draw smaller squares whose sides are the successive Fibonacci numbers 1, 1, 2, 3, 5, 8, 13, 21 and then we can draw a spiral as shown in the diagram below (Petkovic 2009: 15):

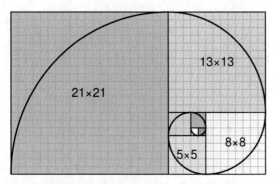

The golden rectangle, Fibonacci numbers, and the spiral form
SOURCE: WIKIMEDIA COMMONS

This spiral form has been found in sunflower heads, pineapples, pinecones, nautilus shells, and many other natural formations. As the reconfiguration of the golden rectangle above suggests, the spiral is an archetype itself that is connected to other archetypes such as iteration and infinity. This connectivity between the golden ratio, the Fibonacci Sequence, and nature is what a hermeneutic geometrician would look for and document via archetypal analysis.

This method can be similarly extended to investigate the appearance of geometric ideas in the human world, such as in art and architecture, and as a framework for understanding some engineering problems. Consider a well-known, yet instructive, example. Engineers often use the triangle form to make bridges stronger. The reason is that four-sided shapes, such as squares and rectangles, can be bent and deformed by forces such as the wind:

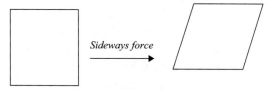

Deformation from a sideways force

However, by adding a beam across the middle of a square, it is turned into two triangles, making it much stronger, since the sides cannot be moved without breaking the beam.

Solution to the deformation effect

By studying such geometrical forms and relating them to physical reality, engineers develop insights that will suggest to them how to make bridges stronger. The same kind of account could be given for many other engineering and architectural practices. Domes, for example, act like arches, which are geometrically strong shapes. In building the dome for the great cathedral in Florence in the Renaissance, Santa Maria del Fiore, Filippo Brunelleschi (1377–1446) arranged the bricks so that the stresses were transferred sideways to the wooden struts in the arch form. Any other arrangement would have allowed the force of gravity to act downwards, forcing the bricks to fall down. Below is the dome:

Brunelleschi's dome

The gist of the foregoing discussion has been to suggest that geometry is both a branch of mathematics and a hermeneutic tool for interpreting the world. Once the work of the geometer is done, the hermeneutician steps in to interpret it semiotically, psychologically, philosophically, and anthropologically. Hermeneutic geometry started when the Pythagoreans aimed to understand the properties of numbers in relation to figures and how these mirrored structures in music and in the movement of celestial bodies. This allowed them to turn practical and intuitive knowledge into powerful philosophical knowledge. It was arguably this power of geometry that impelled Pythagoras to found a secret society to study the *musica universalis*, the connection among geometry, mathematics, music, and the cosmos.

It is of relevance to note that Pythagoras encouraged women to participate fully in his society. Late in life, he married one of his students, Theano. An accomplished cosmologist and healer, Theano headed the Pythagorean cult after her husband's death, and even though she faced persecution, continued to spread Pythagorean hermeneutic philosophy throughout Egypt and Greece alongside her daughters (Pomeroy 2013). A basic tenet of the Pythagoreans was that each natural number stood for something metaphysical. They claimed, for instance, that the number 1 stood for unity, reason and creation. This is why they believed that the single horn of the unicorn possessed magical powers. In the form of a cordial, it continues to have this meaning in many cultures, where it is purported to be able to cure diseases as well as to neutralize the poisons of snakes and rabid dogs.

The Pythagoreans saw geometry as revealing and explaining the inherent Order in the universe. However, they unwittingly discovered irrational numbers, which, like the dissonant ratios in music, were incompatible with their worldview (Maor 2007). At some point, the Pythagoreans noticed that their very own theorem about right-angled triangles, when applied to an isosceles right-triangle with its two sides equal to 1 (the unit length), produced a very strange number for the length of the hypotenuse—$\sqrt{2}$:

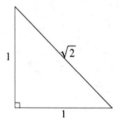

The discovery of $\sqrt{2}$

This was completely unexpected, but there it was—the length of the hypotenuse. The paradox was that $\sqrt{2}$ cannot be written as a *ratio*, since it is a nonrepeating decimal (1.41421356 ...). For this reason, it came to be called *irrational*. Now, this type of serendipitous discovery leads to an age-old debate about mathematics. Do we discover mathematics or do we invent it and then discover that it works? Was $\sqrt{2}$ "out there" ready to be discovered, or did the Pythagoreans produce it inadvertently by their own procedures? Plato believed that mathematical truths pre-existed in the world and that we discover them one at a time. Just like the sculptor takes a clump of marble and gives it the form of a human body, so too mathematicians take a clump of reality and give it symbolic form. In both "representations" we discover many more things about the body and about mathematics. The truth is already in the clump; it takes the sculptor and the mathematician respectively to give it form. The Platonic view has its critics, who point out that we construct mathematical ideas to tell us what we want to know about the world. But, as Berlinski (2013: 13) suggests, the Platonic view is not so easily dismissible:

> If the Platonic forms are difficult to accept, they are impossible to avoid. There is no escaping them. Mathematicians often draw a distinction

between concrete and abstract models of Euclidean geometry. In the abstract models of Euclidean geometry, shapes enjoy a pure Platonic existence. The concrete models are in the physical world.

There might also be a neurological basis to the Platonic view. As neuroscientist Pierre Changeux (2013: 13) muses, Plato's trinity of the Good (the aspects of reality that serve human needs), the True (what reality is), and the Beautiful (the aspects of reality that we see as pleasing) is arguably consistent with notions of modern-day neuroscience:

> So, we shall take a neurobiological approach to our discussion of the three universal questions of the natural world, as defined by Plato and by Socrates through him in his *Dialogues*. He saw the Good, the True, and the Beautiful as independent, celestial essences of Ideas, but so intertwined as to be inseparable ... within the characteristic features of the human brain's neuronal organization.

Plato's model means, however, that we never should find faults within mathematics. As it turns out, this is what Kurt Gödel's (1931) undecidability-incompleteness theorem implied (as will be discussed subsequently). But then, if mathematics is faulty, why does it lead to demonstrable discoveries? René Thom (1975, 2010) referred to these as "catastrophes" in the sense of events that subvert or overturn existing knowledge (Wildgen and Brandt 2010). One of these catastrophes was, clearly, the discovery of $\sqrt{2}$. Thom names the process of discovery as "semiogenesis" or the emergence of meaning-suggestive forms within mathematics. These crystallize by happenstance as mathematicians play around with ideas. As this type of mental play goes on, every once in a while, a catastrophe occurs that leads to new insights, disrupting some of the pre-existing principles within mathematical systems. Discovery is thus catastrophic, in this sense, but this does not tell us why catastrophes occur in the first place. Perhaps the connection between mathematics and reality will always remain a mystery, since the brain cannot really know itself.

Consider a relevant and well-known episode in the history of mathematics, which concerns the Alexandrian geometer Pappus, who was apparently contemplating the following problem: What is the most efficient way to tile a surface? There are three ways to do so with regular polygons—with equilateral triangles, equal four-sided figures, or regular hexagons (Flood and Wilson 2011: 36):

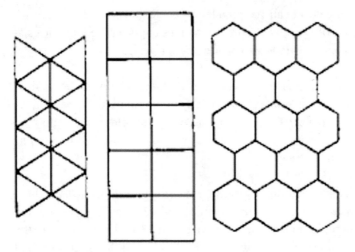

Pappus' floor tiling patterns

The hexagonal pattern has the most area coverage. It is relevant to note that it is a geometric archetype as defined here. Incredibly, bees instinctively use the hexagon as well for their honeycombs, which is the pattern with the most angles, and thus with the capacity for holding more honey than other geometric figures. This is a truly intriguing fact, raising a whole series of hermeneutic questions that take us back to Jakob von Uexküll's (1909) idea that the internal modeling system of a species, such as bees, is well adapted to acting upon the external world. Interestingly, German astronomer Johannes Kepler (1571–1630) made the following relevant observation (cited in Banks 1999: 19):

> What purpose had God in putting these canons of architecture into the bees? Three possibilities can be imagined. The hexagon is the roomiest of the three plane-filling figures (triangle, square, hexagon); the hexagon best suits the tender bodies of the bees; also labour is saved in making walls which are shared by two; labour would be wasted in making circular cells with gaps.

Hexagonal structure also occurs in the molecular configuration of snowflakes and ice crystals. It is little wonder, as Banks (1993: 19) goes on to note, "that currently mathematicians and scientists are devoting much more attention to research on topics advocated by Kepler." Perhaps this shows that each species is tuned into the structure of the world instinctively. In the case of humans, the ability to model this structure geometrically in a conscious, rather than purely instinctive, way is, arguably, what makes them unique among species.

6 Epilogue

Stuart Isacoff (2003) argues that the invention of western musical traditions came about from an unconscious cultural reification of the Pythagorean discovery that the structure of harmony reflects inherent mathematical principles and laws. Thus, the secret to music and why we react so emotionally to it is number. The "re-discovery" of this notion in the early Baroque period dovetailed with the view at the time that the universe operated in accordance with the Pythagorean theory of harmonious Order, as Ian Stewart (2008: 9) aptly remarks:

> The main empirical support for the Pythagorean concept of a numerical universe comes from music, where they had noticed some remarkable connections between harmonious sounds and simple numerical facts. Using simple experiments they discovered that if a plucked string produces a note with a particular pitch, then a string half as long produces an extremely harmonious note, now called the octave. A string two-thirds as long produces the next most harmonious note, and one three-quarters as long also produces a harmonious note. These two numerical aspects of music are traced to the physics of vibrating strings, which move in patterns of waves. The number of waves that can fit into a given length of string is a whole number, and these whole numbers determine the simple numerical ratios. If the numbers do not form a simple ratio then the corresponding notes interfere with each other, forming discordant 'beats' which are unpleasant to the ear. The full story is more complex, involving what the brain is accustomed to, but there is a definite physical rationale behind the Pythagorean discovery.

Indirect support for the Pythagorean discovery came serendipitously from chemistry. In 1865, English chemist John Newlands found that when the elements are arranged according to atomic weight, those with similar properties occur at every eighth element like the octaves of music. Newlands called it, appropriately, the Law of Octaves, and it led to the development of the Periodic Law of chemical elements. This discovery strongly suggests that the structure of matter and music might be somehow interrelated as Pythagoras certainly believed.

But, the Pythagorean paradigm also poses a deep riddle. When ratios between certain string vibrations are set, other ratios are thrown off, making strict use of a single tuning system impossible because it produces dissonances. This means that a fatal defect haunts the Pythagorean model of harmony, which the Pythagoreans knew but kept secret. To banish the dissonances, the keyboard was tempered by breaking the octave into equal parts, so that all harmonies sounded in tune. The most prominent example of this is Johann Sebastian Bach's *Well-Tempered Clavier* (1742). This made the flourishing of western music a reality. In other words, it was a human invention that turned the Pythagorean paradigm into a viable cultural tradition.

An ancient Persian fairy tale, *The Three Princes of Serendip*, may shed some light on the enigma of such discoveries and related inventions. Three princes from Ceylon were journeying in a strange land when they came upon a man looking for his lost camel. The princes had never seen the animal, but asked the owner: Was it missing a tooth? Was it blind in one eye? Was it lame? Was it laden with butter on one side and honey on the other? Was a pregnant woman riding it? Incredibly, the answer to all their questions was "yes." The owner accused the princes of having stolen the animal since, clearly, they could not have had such precise knowledge. But the princes merely pointed out that they had observed the road, noticing several patterns in it: for example, the grass on both sides was uneven, suggesting a lame gait. There were places on the road with chunks of food, like the type that would be produced by a camel with a gap in the mouth chewing food. They also noticed uneven patterns of footprints, the signs of awkward mounting and dismounting typical of someone who was pregnant. Finally, they noticed that there were differing accumulations of ants and flies, which, they explained, would congregate around butter and honey. Their correct questions were thus prompted by inferences (abductions) based on these observations. Knowing the world in which they lived, they were able to make concrete connections between the observations and what happened.

English writer Horace Walpole came across the tale and, since Serendip was Ceylon's ancient name, coined the word *serendipity* to designate how we often come about our discoveries. The term has been applied extensively in science to explain how, for example, Wilhelm Conrad Roentgen stumbled upon what came to be called X-rays by seeing their effects on photographic plates, or Alexander Fleming penicillin by noticing the effects of a mold on bacterial cultures (Roberts 1989, Eco 1998, Merton and Barber 2003). Italian philosopher Giambattista Vico (1668–1744) (Bergin and Fisch 1984), saw the source of serendipitous discoveries in the human *fantasia* (the creative imagination), which he claimed had liberated the human species from the constraints imposed on other species by their particular biology. As Verene (1981: 101) puts it, the imagination allows humans "to know from the inside" by extending "what is made to appear from sensation beyond the unit of its appearance and to have it enter into connection with all else that is made by the mind from sensation." This inner knowing is the reason why we have the capacity to connect things within the mind, beyond the constraints of biology.

Pi in Mathematics and the Physical World

> There is geometry in the humming of the strings, there is music in
> the spacing of the spheres.
>
> PYTHAGORAS (C. 580–500 BCE)

∵

1 Prologue

In a hermeneutic framework, geometry is envisioned as (1) a science of geometrical forms (squares, circles, etc.) and their relation to other parts of mathematics, and (2) an interpretive tool for analyzing the appearance of geometrical ideas and symbols in the natural and physical worlds, such as their
manifestations in music, the movement of the celestial spheres, etc., and (3) a
basis for explaining the use of geometry in human artifacts and creations. In
this paradigm, π would thus be considered to be: (1) an object of geometry
proper; (2) a means to assess the archetypal structures in physical and natural reality; and (3) a part of art, music, and architecture, among other human
practices.

The appearance of π in mathematical formulas describing natural phenomena suggests that the world may have a unified structure and that the brain is
designed to grasp it in bits and pieces as represented by our signs and symbols
(Lotman 1991). So, for example, the manifestation of the golden ratio, ϕ, in the
spiral forms of natural phenomena and the appearance of π in formulas describing natural formations such as waves would fall directly under the rubric
of hermeneutic geometry. This type of approach to knowledge is in line with
the "anthropic principle," as it is called, which implies that we are part of the
world in which we live and thus privileged, in a way, to understand it best. Al-
Khalili (2012: 218) puts it as follows:

> The anthropic principle seems to be saying that our very existence deter-
> mines certain properties of the Universe, because if they were any differ-
> ent we would not be here to question them.

This principle may explain, in part, why π crops up everywhere—in nature and human affairs. As Banks (1999: 183) puts it: "Over the centuries, just about everything involving engineering, science, and technology—from Egyptian pyramids and medieval cathedrals to supersonic aircraft and maglev trains—has utilized mathematics to one extent or another."

This chapter will deal with functions (1) and (2) of hermeneutic geometry— specifically with the appearance of π in other areas of mathematics and in nature. It will start with a brief discussion of Pythagoreanism, and then discuss the manifestations of π in orbital movements, equations describing the DNA double helix, rainbows, ripples from raindrops, wave structures, navigation systems, and the like. Evidence in biology suggests that the rudiments of mathematics may be anchored in our genes, given the fact that infants have an instinctive capacity for recognizing and distinguishing numerical and geometrical concepts (Devlin 2005). However, as suggested in this book, mathematical discovery is largely autopoietic, not fixed by biology. As Charles Peirce (1931–1958, VI: 478) put it, the human mind has "a natural bent in accordance with nature."

2   Pythagoreanism

Pythagoras's ideas and discoveries were known broadly in ancient Greece, influencing Plato, Aristotle, and, through them, subsequent mathematicians and philosophers. Little is known about his life, and most of what is known is more legend than fact. Most historians of science agree that, around 530 BCE, Pythagoras settled in Crotone, in Calabria, where he founded a school, known as the "semicircle," in which members were sworn to secrecy and expected to live ascetically and to be devoted to the study of mathematics (Riedwig 2005: 10). The members of the semicircle came into conflict with the people of Crotone, after the city's victory over Sybaris in 510 BCE. As a consequence, their meeting place was burned down and many were killed. It is not known if Pythagoras himself was among the victims or if he escaped to Metapontum, where, according to some accounts, he eventually died (Pomeroy 2013).

Pythagoras's doctrine of the *musica universalis* is a hermeneutic construct. It implies that a complex structure, such as the universe, is mirrored in a small or representative part of it, such as in music. So what might appear to be coincidences are, upon further analysis, real connections. It is relevant to note that this view is consistent with Carl Jung's (1972) notion of *synchronicity*, which he defined as the simultaneous occurrence of events that appear related and for which we develop explanatory frameworks for connecting them causally. As Jung (1972: 91) emphasized: "When coincidences pile up in this way, one cannot help being impressed by them—for the greater the number of terms in such a series, or the more unusual its character, the more improbable it becomes." So, for instance, the movement of the celestial bodies and the consonant harmonies of music are *synchronic*, that is, they are co-occurrent and might reflect principles of structure symbolized by specific ratios.

Pythagoras is credited with several key discoveries that established mathematics as an autonomous discipline and a scientific language for early astronomy. These include his theorem about right triangles, the discovery that musical tuning is based on specific ratios, the mathematics of the five regular solids, the theory of proportions, the sphericity of the Earth, and the discovery of Venus. He was also one of the first to call himself a philosopher ("lover of wisdom") and the first to classify the Earth into five climatic zones (Hermann 2005). No written work by Pythagoras has survived, likely because oral dialogue was the main mode of interaction among members of the semicircle and because secrecy was imperative to him—which could only be truly guaranteed by avoiding writing discoveries down. Pythagoras's ideas are recounted by several of his contemporaries, such as Philolaus of Crotone. Scattered biographies of Pythagoras from antiquity have also survived, but they are highly unreliable, since they revolve around legends and myths, rather than verifiable historical episodes (Burkert 1972, Kahn 2001, Riedweg 2005).

The *musica universalis* model of a synchronic universe is the core of Pythagoreanism; it was formulated after Pythagoras established that the pitch of a note in music is in inverse proportion to the length of the string that is plucked; also, the intervals between harmonious sounds can be expressed in terms of numerical ratios (Weiss and Taruskin 2008). Pythagoras proposed that the orbits of the sun, moon and planets produce the same ratios found in music. This model of the cosmos was controversial from the outset. Although Aristotle was sympathetic overall to Pythagoreanism, he was highly critical of the *musica universalis* doctrine (Aristotle 350 BCE):

> From all this it is clear that the theory that the movement of the stars produces a harmony, i.e. that the sounds they make are concordant, in spite of the grace and originality with which it has been stated, is nevertheless untrue. Some thinkers suppose that the motion of bodies of that size must produce a noise, since on our earth the motion of bodies far inferior in size and in speed of movement has that effect. Also, when the sun and the moon, they say, and all the stars, so great in number and in size, are moving with so rapid a motion, how should they not produce a sound immensely great? Starting from this argument and from the observation that their speeds, as measured by their distances, are in the same ratios as musical concordances,

they assert that the sound given forth by the circular movement of the stars is a harmony. Since, however, it appears unaccountable that we should not hear this music, they explain this by saying that the sound is in our ears from the very moment of birth and is thus indistinguishable from its contrary silence, since sound and silence are discriminated by mutual contrast. What happens to men, then, is just what happens to coppersmiths, who are so accustomed to the noise of the smithy that it makes no difference to them. But, as we said before, melodious and poetical as the theory is, it cannot be a true account of the facts. There is not only the absurdity of our hearing nothing, the ground of which they try to remove, but also the fact that no effect other than sensitive is produced upon us. Excessive noises, we know, shatter the solid bodies even of inanimate things: the noise of thunder, for instance, splits rocks and the strongest of bodies. But if the moving bodies are so great, and the sound which penetrates to us is proportionate to their size, that sound must needs reach us in an intensity many times that of thunder, and the force of its action must be immense.

Despite critiques of the doctrine, what remains plausible is its incorporation of the principle above—namely, that the structure of the macrocosm is mirrored in the microcosm. This would explain hypothetically why physical and biological structures often display mathematical similarities, as will be further elaborated in this chapter. The Pythagorean doctrine has also been attractive to artists, musicians, philosophers, and even educators across time because it implies connectivity at all levels of human expression and understanding. Medieval education was, for example, based on the Quadrivium, which included arithmetic, geometry, music, and astronomy, which, along with the Trivium (grammar, logic, and rhetoric), made up the seven liberal arts. Contemporary artists and musicians, from Paul Hindemith to the Moody Blues, have incorporated the doctrine into their works, some of which employ musical ratios that are based on Pythagorean principles. So, the *musica universalis* perspective is hardly dismissible, as Aristotle would have it.

3   Uniting Arithmetic and Geometry

Pythagoras was one of the first to unite arithmetic and geometry into an autonomous discipline (mathematics). Even though his mathematical ideas and discoveries are well known, it is worthwhile revisiting some of these here since they are themselves intrinsically hermeneutic. One of these is the concept of *figurate numbers*—numbers that can be displayed geometrically. For example, square integers, such as 1^2 (= 1), 2^2 (= 4), 3^2 (= 9), and 4^2 (= 16) can be arranged in the form of squares:

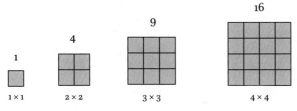

Square numbers

Now, this view of square numbers led to the discovery of an arithmetical pattern that may not have been as obvious otherwise. The pattern is: to form each new square number a successive odd number of little internal squares must be added to it. So, to form the square with sides of 4 units, 3 smaller squares must be added to it; to form the square with sides of 9 units, 5 smaller squares must be added to that square; and so on.

$1 = 1$
$4 = 1 + 3$
$9 = 1 + 3 + 5$
$16 = 1 + 3 + 5 + 7$
$25 = 1 + 3 + 5 + 7 + 9$
...
$n^2 = 1 + 3 + 5 + 7 + 9 + ...$

The Pythagoreans called this pattern *gnomic* because it involved a partitioning method that looked like the carpenter's gnomon:

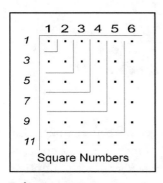

Pythagorean gnomon
SOURCE: WIKIMEDIA COMMONS

The Pythagoreans also came up with the concept of *triangular numbers*. In the diagram below, each dot stands for a numerical unit. So, one dot = the number 1; three dots = the number 3; and so on.

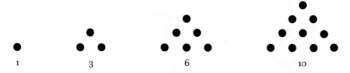

Triangular numbers

The number of dots in the first triangular number is 1; the number in the second is 1 + 2 = 3; the number in the third is 1 + 2 + 3 = 6; and so on. From this, it becomes obvious that each successive triangular number is obtained by adding the integers in order as shown below:

1 = 1
3 = 1 + 2
6 = 1 + 2 + 3
10 = 1 + 2 + 3 + 4
15 = 1 + 2 + 3 + 4 + 5
...
Any triangular number: = 1 + 2 + 3 + ... + n

Now, a comparison between triangular and square numbers reveals yet another pattern—namely that square numbers can be partitioned into triangular numbers as shown in the diagram below:

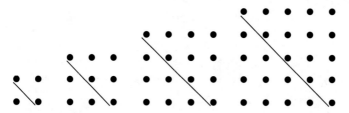

Triangular and square numbers

The Pythagoreans discovered many other connections between numerical and geometric structures. From such constructions they were able to make other key discoveries, such as prime, amicable, and perfect numbers. A prime number is any number that has no factors other than itself and the number 1. The Pythagoreans found that all other integers are made up of a unique set of prime factors—a structural pattern inherent in the integers that was called the

Fundamental Theorem of Arithmetic later by Euclid. Needless to say, the discovery of the primes has had profound implications for mathematics, starting with Euclid who, in the ninth book of the *Elements*, proved that the number of primes is infinite, even though they become scarce as we move up the number line.

Amicable numbers are integers with the following property—the proper divisors of one of them, when added together, produce the other. The numbers 284 and 220 are amicable. The proper divisors of 220 are 1, 2, 4, 5, 10, 11, 20, 22, 44, 55, and 110, and the proper divisors of 284 are 1, 2, 4, 71, and 142. Now, comparing the two we notice the property of "amicability:"

220 = 1 + 2 + 4 + 71 + 142 (the proper divisors of 284)

284 = 1 + 2 + 4 + 5 + 10 + 11 + 20 + 22 + 44 + 55 + 110 (the proper divisors of 220)

A perfect number is one that equals the sum of its proper divisors, with the exception of the number itself. For example, the proper divisors of the number 6 are 1, 2, and 3. Adding these together we get the number 6:

6 = 1 + 2 + 3

The next perfect number is 28. Its proper divisors are 1, 2, 4, 7, 14, and 1 + 2 + 4 + 7 + 14 = 28. Very few perfect numbers have been discovered since Pythagoras's era. Numbers that are not perfect are called excessive or defective. An excessive number is one whose proper divisors, when added together, produce a result that exceeds its value. The number 12, for example, is excessive because the sum of its proper divisors, 1, 2, 3, 4, and 6 (1 + 2 + 3 + 4 + 6 = 16) exceeds its value. A defective number is one whose proper divisors, when added together, produce a result that is smaller than its value. One example is 8, since the sum of its proper divisors 1, 2, and 4 (1 + 2 + 4 = 7) is less than its value.

Now, one can ask if this is nothing more than playing around with number patterns. As it turns out, these numbers have been used in various domains of arithmetical analysis, leading to unexpected discoveries. They have also raised questions about infinity—Are there an infinite number of such numbers? An amicable number did not exist until the Pythagoreans discovered it. This made it possible to use and analyze these numbers to see what they would yield theoretically and practically. From this, other discoveries were made—discoveries that would have been unthinkable without the initial one, since they are derived from it. In sum, the Pythagorean experiments connecting arithmetic and geometry laid the foundations of mathematics as a distinct discipline, showing how certain numbers are related to others and how shapes and number patterns are intertwined.

The Pythagoreans maintained that their discoveries would allow them to understand the mysteries of the universe—a view that was again critiqued by Aristotle, who made the following observation in his *Metaphysics* (1999): "The so-called Pythagoreans, who were the first to take up mathematics, not only advanced this subject, but saturated with it, they fancied that the principles of mathematics were the principles of all things." An example of what Aristotle implied about Pythagoreanism can be seen in the *tetractys*, an equilateral triangle made up of ten dots, with one dot in the top row, two in the second, three in the third, and four in the bottom:

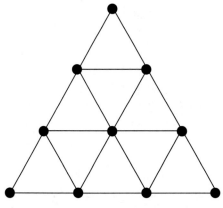

The tetractys

The importance of the tetractys to the Pythagoreans (also called the "mystic tetrad"), is evidenced by the prayer they would recite at meetings of the semicircle (Dantzig 2005: 42):

> Bless us, divine number, thou who generated gods and men! O holy, holy Tetractys, thou that containest the root and source of the eternally flowing creation! For the divine number begins with the profound, pure unity until it comes to the holy four; then it begets the mother of all, the all-comprising, all-bounding, the first-born, the never-swerving, the never-tiring holy ten, the keyholder of all.

The Pythagorean musical system was based on the tetractys given that rows in the diagram, starting at the bottom and going up, can be read as the following musical ratios: 4:3 (perfect fourth), 3:2 (perfect fifth), 2:1 (octave). The use of these ratios has formed the basis of western harmony ever since. The tetractys has had many symbolic meanings. For instance, its four numbers (from the top down), symbolized the overall doctrine of the *musica universalis:*

1 = Unity (the monad)
2 = Dyad (limited/unlimited potential)
3 = Harmony (the triad)
4 = Cosmos (the tetrad)

Since the four numbers in the tetractys add up to ten, it also symbolized a unity of a higher order, namely the Decad, which in music theory designates a set of ten harmonious pitches or pitch-classes. The tetractys also symbolized the organization of space, with the first row representing zero dimensions (a point); the second represented one dimension (a line of two points); the third row two dimensions (a plane defined by a triangle of three points); and the fourth row three dimensions (a tetrahedron defined by four points).

It was such hermeneutic reasoning that greatly influenced Plato, inspiring him to found his Academy in Athens in 387 BCE, in which he emphasized, like Pythagoras, that mathematics provided access to a deep understanding of reality and thus a means for unlocking the secrets of the universe (Fowler 1987). Rather revealingly, the sign Plato affixed above the Academy entrance read: "Let no one ignorant of geometry enter here." Plato saw geometry as a branch of philosophy, astronomy and harmonics. He became known as the "maker of mathematicians," given that his Academy produced some of the most prominent mathematicians of ancient Greece. The fundamental tenet of the Academy was that mathematics and the universe were in harmony—a view that was adopted as well in the Renaissance with the rise of heliocentric theory and a need to revise the ancient models of planetary movements.

4   The Planetary Orbits

In his classic 1619 treatise titled *Harmonices Mundi*, German astronomer Johannes Kepler constructed mathematical equations of orbital movements that connected geometry, astronomy, and musical harmony into a viable model of the universe that has remained to this day. Kepler showed how the structure of musical harmony mirrored the structure of the orbits of the planets in the solar system, thus reviving and assigning scientific justification to the Pythagorean concept of *musica universalis*. Specifically, Kepler demonstrated that the angular velocities of planets measured from the Sun were mirrored in musical intervals.

For example, he found that the maximum and minimum orbital velocities of Saturn (in terms of arc seconds from the Sun) differed by an almost perfect 5:4 ratio (a major third in music); the orbits of Jupiter by a 6:5 ratio (a minor third in music); and the orbits of Mars by 3:2 (a perfect fifth). Kepler also examined the ratios between the fastest or slowest speed of a planet and the slowest or fastest speed of its neighbors (which he called converging and diverging motions), discerning harmonies in each case with discrepancies less than 24:25, the smallest harmonic interval that can be perceived by human hearing. The only interval among the orbits with a larger deviation was the diverging motion of Mars and Jupiter. Writing in 1619, Kepler could not explain this because he did not know of the asteroid belt between the orbits of these two planets, which was discovered later in 1801.

Incorporating the insights of his own teacher, Tycho Brahe (1602), Kepler formulated three laws of planetary motion, which can be paraphrased as follows:

1. The planets move in elliptical orbits with the Sun at one focus.
2. The line from the Sun to a planet covers equal areas during equal time intervals.
3. The square of the period of any planet's orbit is proportional to the cube of the semi-major axis of its orbit.

Kepler derived his first law from his calculations of the orbit of Mars, inferring that all planetary orbits are similarly elliptical, since the laws of nature imply symmetry and consistency, not aberration. The second law establishes that when a planet is closer to the Sun, it will travel faster. The third law indicates that the farther a planet is from the Sun, the longer its orbit, and vice versa. Kepler's laws describe the speed of a planet traveling in an elliptical orbit around the sun, establishing that a line between the Sun and the planet sweeps equal areas in equal times. Therefore, the speed of the planet increases as it nears the Sun and decreases as it recedes from the Sun. A diagram that illustrates Kepler's laws is the following one:

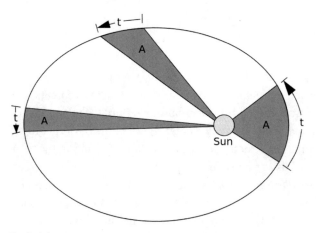

Kepler's laws
SOURCE: WIKIMEDIA COMMONS

Pi appears in the calculations of the second and third laws. Without going into the mathematical details here, this appearance of π is connected to Newton's Law of Gravitation, which states that any two objects attract each other with a force proportional to the product of their masses and inversely proportional to the square of the distance between them. From a hermeneutic perspective, the appearance of π in the calculations is not surprising, since planetary orbits are elliptical and thus based on the archetype of circularity, of which elliptic curvature is a topological derivative. The fact that π appears in calculating planetary orbits and that Kepler was able to show how these were connected to musical harmonies is remarkable nonetheless. It suggests that the structure of the world may be archetypal after all and that we discover this through mathematics.

As Kepler's work showed, the *musica universalis* doctrine had survived the transition from a geocentric to a heliocentric model of the Solar System. Kepler's scientific justification of the Pythagorean system implied that archetypal structure may indeed be built into the cosmos.

5 Natural and Physical Phenomena

To encapsulate the foregoing discussion, the appearance of π in the mathematical description of planetary orbits is due to their elliptical motion—a topological derivative of circularity. This suggests that circularity manifests itself in various shapes and forms possessing the more general archetype of curvature, with the circle being a specific manifestation of curvature. The fundamental properties of circularity are thus unaffected by the topological transformation of circularity, hence the appearance of π in the mathematics of curved shapes such as ellipses.

Another hermeneutic principle related to circularity is, as discussed in the previous chapter, its connection to other archetypal structures, such as infinity, probability, recursion, iteration (oscillation, periodicity, etc.). This may plausibly explain why π appears in formulas that describe rhythms and waves, which have iterative or periodic structure. It surfaces, for example, in several sine and cosine functions from trigonometry that repeat their values in regular intervals or periods. An example is the sine function, which repeats every 2π radians (or $360°$):

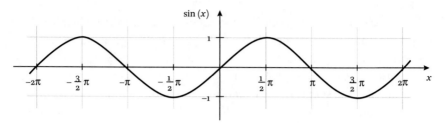

Sine function
SOURCE: WIKIMEDIA COMMONS

Periodic functions describe oscillations, waves, and other phenomena that exhibit periodicity. Significantly, π appears in the Fourier series. This is an infinite series of trigonometric periodic functions, named after Jean-Baptiste Joseph Fourier (1768–1830). It has been used to describe all kinds of periodic phenomena, such as the rhythms of infant breathing, the circadian rhythms of sleep, among others. Pi surfaces regularly in functions that describe the ebb and flow of the ocean's tides, the electromagnetic waves in telecommunications, and the behavior of atoms and subatomic particles. In essence, π is woven into the structure of natural and physical matter wherever curvature, rhythm, periodicity, and other derivatives of circularity are involved. Agarwal, Agarwal, and Syamal (2013: 2) summarize this general finding as follows:

> Since the exact date of birth of π is unknown, one could imagine that π existed before the universe came into being and will exist after the universe is gone. Its appearance in the disks of the Moon and the Sun, makes it as one of the most ancient numbers known to humanity. It keeps on popping up inside as well as outside the scientific community, for example, in many formulas in geometry and trigonometry, physics, complex analysis, cosmology, number theory, general relativity, navigation, genetic engineering, statistics, fractals, thermodynamics, mechanics, and electromagnetism. Pi hides in the rainbow, and sits in the pupil of the eye, and when a raindrop falls into water π emerges in the spreading rings. Pi can be found in waves and ripples and spectra of all kinds and, therefore, π occurs in colors and music. The double helix of DNA revolves around π. Pi has lately turned up in super-strings, the hypothetical loops of energy vibrating inside subatomic particles.

Like the sine function above, the cosine function also shows periodicity; its wave structure is symmetrical to the sine function, but the two functions are not coincident, crisscrossing at regular intervals, as can be seen when they are plotted together:

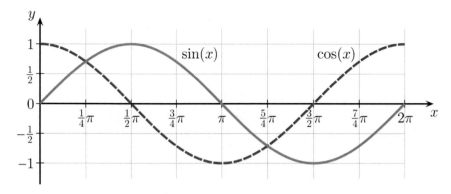

Sine and cosine functions

As it turns out, these functions can be defined archetypally in relation to the "unit circle," a circle with a radius equal to 1. When we plot the unit circle on the Cartesian plane, we can see why this is so by drawing a triangle within it, as shown in the diagram below:

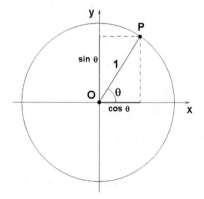

Unit circle

This shows that these trigonometric functions are linked structurally to the circle and, thus, to the circularity archetype. In turn, this would explain why π is embedded in them and, by extension, why periodic and oscillatory phenomena represented by these functions involve π.

The tangent function (tan) produces a different kind of graph: it moves between negative and positive infinity. At ±π/2 radians, ±3π/2 radians, etc. the function is undefined:

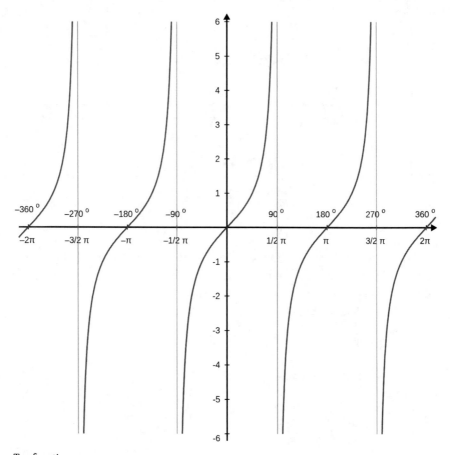

Tan function
SOURCE: WIKIMEDIA COMMONS

Mathematical models such as the ones above allow the scientist to describe phenomena such as waves, periodic phenomena, the shape of rivers, etc. A river's movements and shapes are determined by its "meandering ratio;" this has been shown to approach π. Einstein (1926) used fluid dynamics and chaos theory to show that rivers tend to bend into loops. The reason is that the slightest curve in a river will generate faster currents on the outer side of the curve, which will cause erosion and a sharper bend—a process that will gradually form a loop, until chaos causes the river to suddenly double back on itself, at which point it will begin forming a loop in the other direction. This is an extraordinary finding, as Simon Singh (1997) indicates:

> The number π was originally derived from the geometry of circles, and yet it reappears over and over again in a variety of scientific circumstances. In

the case of the river ratio, the appearance of π is the result of a battle be-tween order and chaos. Einstein was the first to suggest that rivers have a tendency toward an ever more loopy path because the slightest curve will lead to faster currents on the outer side, which will in turn result in more erosion and a sharper bend. The sharper the bend, the faster the currents on the outer edge, the more the erosion, the more the river will twist, and so on. However, there is a natural process that will curtail the cha-os: increasing loopiness will result in rivers doubling back on themselves and effectively short-circuiting. The river will become straighter and the loop will be left to one side, forming an oxbow lake. The balance between these two opposing factors leads to an average ratio of π between the ac-tual length and the direct distance between source and mouth.

In the domain of physics, one of the most important appearances of π is, actually, in one of Einstein's field equations relating the curvature of the space-time con-tinuum to energy (Schmidt 2000). Without going into the mathematical details here, suffice it to say that the relevant formula calculates how objects with a large mass, such as stars and galaxies, can curve space and time with their gravity. The formula can be simplified as follows, where G is Newton's constant of gravitation:

Gravity = 8πG (Energy + Momentum)

Einstein figured out that gravity can be described by connecting fields (regions in which points are affected by a force, such as when objects fall to the ground because they are affected by the force of Earth's gravitational field) to localized bodies on the surface of a sphere, which reveals the archetypal reason why π appears in the formula.

Pi also surfaces in mathematical descriptions of organic processes. It ap-pears, for instance, in morphogenesis—the process that underlies how an organism derives its shape or form. In animals, the embryo develops from a uniform group of cells which generate a brain, backbone, and limbs. In 1952, Alan Turing, developed a mathematical model for morphogenesis. An embryo is shaped into different anatomical patterns by chemicals termed "morpho-gens" by Turing, which spread through tissues. In the simplest scenario, a pat-tern results from the reaction of two morphogens, an activator and an inhibi-tor. The former self-amplifies and can only spread locally. But in the process it stimulates the growth of the inhibitor which, in turn, suppresses the activator, and spreads considerably. Computer models of this system reveal that Turing's model produces an array of spots and stripes (Davies 2013). The activator mor-phogen forms local patches of spots or stripes, while the inhibitor morphogen prevents the patches from growing too close to each other. Turing's model is now employed to explain the formation of animal fur coats, pigment shapes in tissues, and limb structure. It is relevant to note that π appears in the mathematical description of several aspects of morphogenesis, such as the size and spacing of many patterns, including cell division timing, heart beats, breathing cycles, and circadian rhythms controlling sleep-wake cycles.

The manifestations of π in all kinds of physical and biological systems is suggestive that circularity is a generic archetype, or "meta-archetype," and may thus be part of the fabric of existence. Pi is a trace to circularity and its topological derivations—periodicity, curvature, sphericity, etc.—in physical and biological structure. As another case-in-point, let's briefly revisit Buffon's Needle Problem (Chapter 1), which is explained in terms of a normal probability curve, thus entailing π because the area under the curve, known as its integral in calculus, is equal to the square root of π. Buffon himself noted that two times the number of needles dropped divided by the number of needles crossing a line was almost equal to π. He used these results to provide his own estimate of the value of π. A graph showing how the probability of the needle crossing the line might occur is the following one—notice that the graph goes on ad infinitum to the right, thus linking π in the needle experiment to the archetype of infinity.

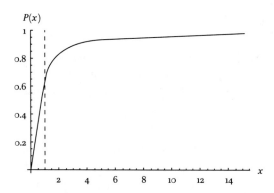

Graph for Buffon's needle problem
SOURCE: WOLFRAM MATHWORLD

The task of any science is to explain how and why phenomena are the way they are by means of models or theories. As new facts emerge or are collected about the relevant phenomena, science will adjust, modify, or even discard previous models on the basis of the new data. In this cumulative fashion, the ultimate goal is to explain what Aristotle called the "final causes" of reality. But it seems that all we end up having are the models themselves, but not any true explanation of why reality is the way it is. The movie π: Faith in Chaos, as discussed several times, treats this conundrum brilliantly, providing its own answer—namely that if we were to unravel the hidden pattern in the digits of π, perhaps we could grasp reality in its essence—a theme to be discussed in the final chapter.

6 Topology, Non-Euclidean Geometry, and Fractal Geometry

The Pythagoreans were averse to dissonance and chaos. The paradox of π is that it is a number that appears to be random in its digit constitution, yet is derived from one of the most perfect (consonant) of all figures—the circle and its topological derivatives (cycles, spheres, etc.). Given this perspective, it is unlikely that they would have been interested in the properties of deformations of geometric figures. Interest in these had to await the advent of topology in the eighteenth century, which studies properties of shapes which do not change after transformation. As discussed in the previous chapter, a circle will always divide the plane into an area inside the circle itself and the area outside, no matter how it is transformed, such as into a crinkly curve. That is a property of circularity and thus applies to spheres and other forms that have an inside and an outside. Topology is thus a powerful hermeneutic system for explaining why π crystallizes in formulas that seem, at first, to have nothing to do with circles.

To illustrate concretely how topological properties are established, imagine a doughnut made of soft clay. By stretching and pinching, but not tearing or joining anything, the doughnut can be molded into a coffee mug. But the hole in the original doughnut shape remains—as the hole in the handle of the coffee mug. It is thus invariant. A drawing of the transformation from doughnut to coffee mug was shown by Stephen Barr in his fascinating book, *Experiments in Topology* (1964):

Transforming a doughnut into a coffee mug
SOURCE: FROM BARR 1964

Because they share the property of invariance, the doughnut and coffee mug are topologically equivalent. Concepts such as invariance and equivalence are at the core of topology as a modeling device of reality. In this framework, a circle is topologically equivalent to an ellipse (it just needs stretching) and a sphere is topologically equivalent to a spheroid—a surface obtained by rotating an ellipse about one of its principal axes. The first systematic treatment of topology is found in a 1847 article by the nineteenth-century German mathematician Johann Benedict Listing (re-issued as a monograph in 1848), who first used the term *topology*. However, the underlying ideas go back to Leibniz in the seventeenth century and Euler in the eighteenth.

The relevant point here is that topology has become a key to unlocking natural structures, such as the DNA. Stewart (2012: 105) elaborates its importance as follows:

> One of the most fascinating applications of topology is its growing use in biology, helping us understand the workings of the molecule of life, DNA. Topology turns up because DNA is a double helix, like two spiral staircases winding around each other. The two strands are intricately intertwined, and important biological processes, in particular the way a cell copies its DNA when it divides, have to take account of this complex topology.

Topological ideas not only model physical phenomena, but also suggest hidden or emergent structural possibilities. So-called manifold topology has become significant in geometry and physics, because it makes it possible to describe complicated structures in terms of the simpler local topological properties of Euclidean geometrical spaces. Manifolds arise as solution sets of systems of equations and as graphs of functions.

Non-Euclidean geometries emerged at about the same time as topology (Coxeter 1942). In these, the geometric figures studied are not limited to Euclidean spaces and principles, contravening Euclid's Parallel Postulate that only one line through a given point can be parallel to another given line. During Euclid's time, and for centuries thereafter, mathematicians attempted to prove that the postulate could be derived from Euclid's other axioms. Now, in two-dimensional space (the plane), parallel lines do go on indefinitely without ever crossing, like train tracks; but this does not hold in different dimensions or different surfaces.

In the 1800s, mathematicians finally showed that the postulate cannot be proved from the other axioms and does not hold in different topological spaces. From this, non-Euclidean geometries emerged. In one of these, called hyperbolic or Lobachevskyan geometry (after Nikolai Lobachevsky 1855), the parallel postulate is shown not to hold with regard to straight lines within the circle where, it can be shown that through a point, *P*, many lines can be drawn across the circumference which can be defined as parallel to another line, *a*, that does not go through that point, as shown below—"parallel" understood as not-crossing *a*:

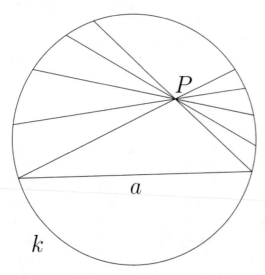

Lines within the circle

Around 1860, Bernhard Riemann extended the notion of "parallel" to there-dimensional space. On a sphere, such as the globe, the lines drawn across it are great circles. As can be seen in the diagram below, these crisscross; but at the poles they are parallel.

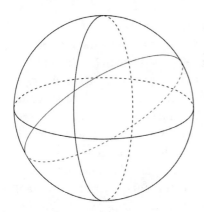

Great circles on a sphere

The foregoing discussion of non-Euclidean geometries is a reductive one; the objective has been simply to emphasize that geometry has been extended to describe non-Euclidean spaces and figures. As a hermeneutic science, geometry has been used since antiquity to describe the physical world; but as our knowledge of the world changes, so too does our geometric modeling of the world. So, we might ask: Which type of geometry, Euclidean or non-Euclidean, provides the best model of reality? Some situations are better described in non-Euclidean terms, such as aspects of the theory of relativity. Other situations, such as those related to everyday building, engineering, and surveying, seem better described by Euclidean geometry. In other words, geometry is situation-specific.

Non-Euclidean geometries have led to many scientific discoveries that would not have been thinkable without them. To quote Richeson (2008: 201):

> But even the most abstract and theoretical areas of mathematics can prove useful. Applicable mathematics often comes from decidedly non-applicable areas. The usefulness of a particular theory is often not clear for many years. No one could have predicted that the study of prime numbers would later enable us to encrypt credit-card information so that it can be sent safely across the Internet. Nineteenth-century mathematicians did not know that their work in non-Euclidean geometry would provide the foundation for Einstein's theory of general relativity. Toward the end of the nineteenth century the usefulness of knot theory reemerged in the natural sciences. Physicists, biologists, and chemists discovered that the mathematical theory of knots gave them insight into their fields. Whether it is the study of DNA or other large molecules, magnetic field lines, quantum field theory, or statistical mechanics, knot theory now plays an important role.

Another type of non-Euclidean geometry that is relevant to the present discussion is *fractal geometry*—a term first used by mathematician Benoit Mandelbrot in the 1970s. Mandelbrot had an abiding interest in self-similarity, a property of geometrical figures that have a similar appearance or structure when viewed at different scales. In his paper "How Long is the Coast of Britain?" published in *Science* magazine in 1967, he discussed the phenomenon of self-similarity in coastlines. Shortly thereafter, in 1975, he used the term *fractal* to describe any repeating self-similar pattern (see Mandelbrot 1982).

Self-similarity refers to shapes that are copies on different scales. An example of a fractal shape is the Mandelbrot set, or M-set, a pattern resembling an ink splash generated by a simple equation. When graphed on a computer at increasing magnifications, the set reveals a series of patterns that are very similar but which differ slightly in detail:

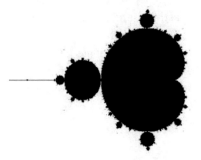

The M-set

The mathematics behind the M-set is relatively simple; it involves the following function (reduced for illustrative purposes):

$$z = z^2 + c$$

Mandelbrot found that for certain values of z the outputs would continue and grow forever, while for others they would shrink to zero. This is shown as follows. Each value of c is plugged back into the function. As we iterate the function in this way, one of two outcomes is possible: either the iterations get larger and larger, moving further and further away from o, or they stay close, never getting very far from o.

Now, an intriguing aspect is that π has been found to be hidden within such iterations. It was computer scientist David Boll who discovered this (see Klebanoff 2001). Pickover (2005: 87) describes Boll's discovery as follows:

> In 1991, David Boll discovered a strange connection between π and the classic Mandelbrot set…, which can be visualized as a bushy object that describes the behavior of $z = z^2 + c$, where z and c are complex numbers. Boll studied points lying on a vertical line through the point $z = (-0.75, 0)$ and discovered that the number of iterations needed for these points to escape a circle of radius 2 (centered at the origin) is related to π. In particular, consider points $z = -0.75 \pm \varepsilon i$ and the n iterations needed for orbits of these points to escape the circle. As ε goes to o, $n\varepsilon$ approaches π.

Formulas for human lungs, trees, clouds, and other biological and physical phenomena turn out to have fractal structure. Fractal geometry has thus emerged as a "secret code" of nature, telling us that iteration is an inherent principle in the structure of the universe, at least in some of its parts. Coined from the Latin word *fractus*, the term fractal suggests fragmented, broken and discontinuous phenomena. But, as it turns out, fractals are apt models for understanding the shapes of fern leaves, snowflakes, lava flows, coastlines, and mountain terrains, revealing their iterative structure.

Consider a well-known fractal form, called the *Sierpinski Triangle*, named after the Polish mathematician Vaclav Sierpinski (Barnsley 1988). It is an equilateral triangle subdivided iteratively inside with upside-down smaller equilateral triangles:

Sierpinski triangle

Remarkably, parts of nature, such as snowflakes, are based on an isomorphic fractal structure—a fact discovered by Swedish mathematician Helge von Koch in 1904. To model the snowflake, Koch took an equilateral triangle modifying it with a sequence of successive changes, each one formed by adding outward bends to each side of the previous triangle, making successively smaller equilateral triangles. The Koch snowflake fractal model is remarkably similar to real snowflakes:

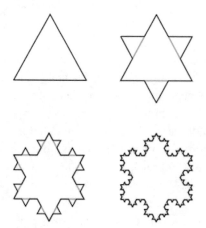

Koch snowflake
SOURCE: WIKIMEDIA COMMONS

In a phrase, fractals are effective models for understanding the shapes of many natural formations. To cite Solomon Marcus (2012: 179):

> What art and poetry anticipated in the 19th century, together with some phenomena pointed out by Weierstrass, Peano and Koch, related to curves devoid of tangents in all their points, became explicit in the mathematics of the second half of the past century, when Benoit Mandelbrot invented *the fractal geometry of nature*. Its idea is that nature, in most of its aspects, is not at all simple and regular. Clouds, ocean coasts, Brownian motion, snowflakes, mountains, rivers don't fit with the regular objects of traditional geometry. Even celestial bodies, long-time considered models of regularity, prove to be less regular than they are supposed to be. How to approach this world of high complexity? The answer proposed by Mandelbrot is the notion of a fractal object. Such objects are obtained as limits of some asymptotic processes, starting with some regular figures. If the first steps of these processes are visible objects and fit with the simplicity of traditional geometry, as soon as we go to the next steps the new objects become less and less visible and regular. At the limit, we get completely invisible, however perfectly are its intelligible objects, the fractal ones. What makes them very attractive is their inner, hidden simplicity, in contrast with their outer complexity: in a fractal object, there is a remarkable phenomenon of self-similarity: it repeats at its different levels in the same structure As a matter of fact, everybody can test this fact looking carefully at the structure of a tree in the forest.

The core ideas that led to Mandelbrot's founding of fractal geometry are traced to a paper presented by Karl Weierstrass to the Royal Prussian Academy of Sciences in 1872 (Edgar 2004: 7–11). In it, Weierstrass provided the first definition of a function with a graph that would today be considered a fractal, since it was a function that was continuous but not differentiable (that is, without derivatives). Shortly after, in 1883, Georg Cantor, published examples of subsets of the real line known as Cantor sets, which had properties that are now recognized as fractals (Edgar 2004: 11–24). By the last part of the nineteenth century, Felix Klein and Henri Poincaré developed these ideas further, culminating in the work of Mandelbrot (1977). But fractal shapes were known long before fractal geometry provided a theory for them. They turn up in art across cultures and across time. In Mahayana Buddhism, the fractal nature of reality is captured in the Avatamsaka Sutra by the god Indra's net, a vast network of precious gems hanging over Indra's palace, so arranged that all the gems are reflected in each one. Another magnificent example is on the main dome of Selimiye Mosque in Edirne, Turkey, crested with self-similar patterns:

Dome of the Selimiye Mosque in Edirne
SOURCE: WIKIMEDIA COMMONS

Examples of fractal art and architecture are found across time and across so-cieties—a fact that lends anecdotal support to iteration and recursion as based on archetypal structure. In recent times, artists Salvador Dalí, Jackson Pollock, and Maurits Escher have exploited the fractal technique (Bovill 1996). Today, fractal art is usually created with fractal-generating software. But this does not mean that it does not involve the creativity of the artist. As Kerry Mitchell (2015) has argued, fractal art is not just computerized art; it requires the same kind of creativity that paintbrush art does. The difference is that instead of the brush, the fractal artist uses the computer. As Mitchell puts it:

> Requiring of input, effort, and intelligence, the Fractal Artist must direct the assembly of the calculation formulas, mappings, coloring schemes, palettes, and their requisite parameters. Each and every element can and will be tweaked, adjusted, aligned, and re-tweaked in the effort to find the right combination. The freedom to manipulate all these facets of a fractal image brings with it the obligation to understand their use and their ef-fects. This understanding requires intelligence and thoughtfulness from the Artist.

7   Epilogue

A fundamental premise of hermeneutic geometry is that archetypal structure is omnipresent in the world, and that geometry allows us to discover it. As the historian of science, Jacob Bronowski (1973: 168) has insightfully written, we hardly recognize today how important the study of geometry has been to the progress of civilization—an evolutionary thrust forward initiated by Pythagoras who established "a fundamental characterization of the space in which we move ... [describing] the exact laws that bind the universe."

There is a commonly-held notion among mathematicians that all mathematics is trivial once it is understood (Hardy 1967). To put it more specifically for the present purposes, once something like the presence of π in scientific descriptions of reality is discovered, we lose our fascination with it. It is easy today to view the fact that π appears in the description of natural and physical phenomena as self-evident. However, if we step back and think about it, it is still amazing that π is a trace to unraveling some of the world's mysteries. But this raises a deep existential question: Why is reality this way? There really is no definitive answer to this, as the movie π: *Faith in Chaos* suggests.

All that can be affirmed is that there is a connection between mathematics, the mind, and reality. *What Is Mathematics?* is the title of a significant book written for the general public by Courant and Robbins in 1941. Their answer to the question in the title was an indirect one—they illustrate what mathematics looks like and what it does. And perhaps this is the only possible way to answer the conundrum of why π is ubiquitous in nature—all we can really do is document its presence. A year before, in 1940, Kasner and Newman published another significant popular book titled *Mathematics and the Imagination*. The authors showed how mathematics is used to probe the meaning of quantity and space in the universe. Although the authors of these books did not use the term *hermeneutic science*, the way they described the relation between mathematics and reality was essentially hermeneutic.

Our understanding of the world is shaped by our sense organs. Certain wavelengths of electromagnetic energy stimulate our eyes, others do not. Our ears can process certain kinds of mechanical vibrations in the air, but ignore others. Our noses and tongues are sensitive to certain kinds of chemical stimuli, but not to others. Sense organs change environmental input into nervous impulses, which then go to the brain, where the sensory information is turned into understanding via modeling processes. Our models of the world thus become stable for a while, until what Thom called a cognitive "catastrophe" emerges (Chapter 1). When this happens, new insights on some phenomenon are gained unexpectedly yet significantly.

A startling example comes from 2015, involving Wallis's formula for π (Chapter 1):

$$\pi/2 = 2/1 \times 2/3 \times 4/3 \times 4/5 \times 6/5 \times \dots$$

Mathematician Tamar Friedman and physicist Carl Hagen (2015) found this formula unexpectedly in the structure of the hydrogen atom. The discovery was made, apparently, while Hagen was teaching a class in quantum mechanics on the so-called "variation principle"—a principle to approximate the energy states of a hydrogen atom (Macdonald 2018). As he was illustrating the conventional calculations, Hagen noticed an unexpected pattern in the ratios. So, he asked mathematician Friedman to help him figure out what was going on. Together, they identified the pattern as a manifestation of the Wallis formula for π, which was the first time it had appeared in quantum physics. Since 1656, when Wallis proposed it, there have been many proofs of the formula, but all had come from the world of mathematics. The fact that it emerged in the field of the quantum mechanics of the hydrogen atom is astonishing. It constitutes one of those "obscure secrets" that Ahmes talked about in the opening part of the *Rhind Papyrus* and which, as Ahmes also indicated, could be at the very least brought to awareness by mathematical reasoning, even if we may have no real explanation for their presence in the world (Chapter 1).

CHAPTER 3

Pi in Art and Architecture

Science discovers; art creates.

JOHN OPIE (1761–1807)

∴

1 Prologue

The Great Pyramid at Giza was built around 2500 BCE. It has a perimeter of approximately 1760 cubits—around 440 cubits per side on its square base—and a height of around 280 cubits. When put into a ratio, 1760/280, the result is approximately 2π.

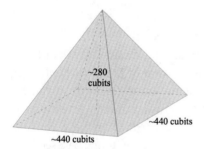

Pi in the Great Pyramid

Some historians believe that this ratio was not chosen randomly, but for symbolic and aesthetic reasons (Verner 2003: 70). Others dispute this (Petrie 1883: 30), claiming that the Egyptians had no concept of π which, as discussed in Chapter 1, is not quite correct (recall Ahmes' proof of π). Whatever the truth, explaining the appearance of π in architectural structures, intentionally or serendipitously, would fall under the rubric of hermeneutic geometry, since one of its central aims is to understand why geometric constructs such as π or φ (the golden ratio) have been incorporated, wittingly or unwittingly, into cultural, symbolic, artistic, and architectural practices and traditions, and then to make connections through them between the natural and human worlds.

The Egyptian builders of the pyramid left no record that they knowingly used π in their construction plan. It was John Taylor, in his 1859 book, *The Great Pyramid: Why Was It Built? And Who Built It?* who first proposed that π was used intentionally in the design of the Great Pyramid. He suggested that the reason for this was that the Great Pyramid was intended to incorporate into its structure the geometric fact that the Earth was a sphere, with the height corresponding to the radius joining the center of the Earth to the North Pole and the perimeter corresponding to the Earth's circumference at the Equator. Taylor also claimed that the golden ratio was also incorporated intentionally into the construction of the Great Pyramid. It is visible in an internal right triangle with hypotenuse φ and sides 1 and √φ:

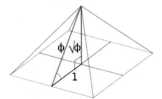

Phi in the Great Pyramid

Taylor's claims have been critiqued on several counts (see Hitchins 2010). One is that the triangle that contains the golden ratio inside the Great Pyramid was not known until it was uncovered much later by Johannes Kepler; for this reason it is called the Kepler Triangle. Nonetheless, the fact that geometric constructs can be detected in the design of ancient structures—intentionally or not—raises the hermeneutic question of why they appear in them. The link between geometry, aesthetics, and symbolism is an ancient one. We seem to have always perceived some geometrical forms not only in terms of their practical utility but also in terms of the aesthetic qualities that they possess. When asked to visualize a triangle, for example, people living in western culture will tend to call to mind the equilateral triangle, perceiving it to be exemplary or representative of the triangle form itself. Obtuse-angled, right-angled, and acute-angled triangles are perceived, instead, to be subtypes. The reason for this reaches back into the meanings of triangles in ancient Greece—meanings carried over into Renaissance geometry and art where symmetry and perfection of form were similarly praised. The equilateral triangle as a cultural prototype is arguably the result of this tradition, working its way into perception through artistic practices.

The research on children's drawings is instructive in this regard. If a drawing instrument is put in a child's hand, that child will instinctively use it to draw—a skill that no one has imparted or transmitted to the child explic-itly. As Dissanayake (1992: 79–80) explains, this tendency reveals that we are inclined since infancy to perceive geometric shapes as pleasurable in themselves:

> If given drawing materials around age two or three, young children happily scribble randomly on the drawing surface. As time passes, how-ever, their scribbling becomes more and more controlled; geometrical shapes such as rude circles, crosses, and rectangles, at first accidental-ly produced, are repeated and gradually perfected. Although children, with parental prompting, may learn to label circles as "suns" or "faces," to begin with they do not set out to draw anything in the environment but instead seem spontaneously to produce forms that become refined through practice into precise, repeatable shapes. The act of making shapes is pleasurable in itself and appears to be intrinsically satisfying; usually identification is provided, if at all, only after the child finishes drawing. Of course, shapes eventually suggest "things" to the child as the ability to use symbols develops, but in the beginning, pleasure and sat-isfaction occur without larger or more explicit associations of meaning. This form of activity in the presymbolic child is perhaps truly an exam-ple of "art for art's sake."

In a relevant study, Villarroel and Ortega (2017) found that similar and even identical geometric shapes emerge in children's drawings, no matter the cul-tural context of rearing. They encapsulate their findings as follows (Villarroel and Ortega 2017: 85):

> Evidence provided by this study is coherent with the assumption that be-fore starting primary education, internal representation of closed curves, quadrilaterals and triangles should have been developed by a significant number of children and, more interestingly, that young children's graph-ical expressivity appears liaised to their skill to depict two-dimensional geometric shapes.

In yet another study, Izard, Pica, Pelke, and Dehaene (2011) looked at notions of Euclidean geometry in an indigenous Amazonian society, where such ge-ometry is not part of upbringing. The research team tested the hypothesis that certain aspects of Euclidean geometry are based on intuitions of space that are present in all humans (such as points, lines, and surfaces), even in the absence of formal mathematical training. The Amazonian society is the

Mundurucu—other subjects in the study included adults and age-matched children controls from the United States and France as well as younger American children without training in geometry. The responses of Mundurucu adults and the children in the study converged with those of mathematically educated subjects, suggesting an intuitive understanding of essential properties of Euclidean geometry. For instance, on a surface described to them as perfectly planar, the Mundurucu's estimations of the internal angles of triangles added up to approximately 180 degrees, and they stated that there exists one single parallel line to any given line through a given point. These intuitions were also present in the group of younger American participants. The researchers concluded that humans instinctively develop geometrical notions that are in accord with the principles of Euclidean geometry, no matter where they are reared.

2 Pythagoreanism in Art

The Pythagoreans not only contributed to our understanding of the universe through hermeneutic geometry, but they also ignited a revolution in Greek art and architecture, as sculptors and architects often used their geometric ideas to create their works (Homann-Wedeking 1968: 62–65). The belief of the artists and architects was that beauty lay in proportion—that is, in the interrelation of the parts with one another and with the whole, like the notes and tones in music. A ratio such as 2:1, for instance, as Rhys Carpenter (1921: 107) has observed, was "the generative ratio of the Doric order, and in Hellenistic times an ordinary Doric colonnade."

Strangely, Plato, who saw geometry as a powerful hermeneutic science for exploring reality, did not link it to aesthetics, perhaps because he viewed art as artificial with no connection to truth and thus having no real philosophical purpose: "the whole art of imitation is busy about a work which is far removed from the truth; and is its mistress and friend for no wholesome or true purpose; is the worthless mistress of a worthless friend, and the parent of a worthless progeny" (Plato 2008). Plato thus saw art as separate from hermeneutic geometry, which he believed should be pursued "for the sake of the knowledge of what eternally exists, and not of what comes for a moment into existence, and then perishes" (Plato 2008). Aristotle (1952) also saw art as imitation, but he did not believe that it was purposeless, but rather that its function was to complete what nature did not finish. Thus, art had the capacity to profoundly affect the human observer and transform the spirit. Aristotle too, however, did not see the principles of hermeneutic geometry and art as interrelated.

Despite such views, a large portion of ancient Greek art and architecture was realized according to these principles (Homann-Wedeking 1968: 63). The Greek artists, like the Pythagoreans, believed that nature expressed it-self in archetypal forms and structures which were mirrored in geometric ratios and figures like the circle, the triangle, and the square. The fifth century BCE sculptor Polykleitos, whose art techniques were based directly on geometric ratios, intended to produce aesthetic perfection via *symmetria* ("symmetry"), *isonoimia* ("equilibrium"), and *rhythmos* ("rhythm"). As Kenneth Clark (1956: 63) has observed, Polykleitos' technique was designed to convey a sense of "clarity, balance, and completeness." As Clark goes on to illustrate, this technique was based on repetition, which, as discussed, is an archetypal process (Chapter 2). So, in sculpting the human body Polykleitos started with a specific body part, such as a little finger, treated as the side of square. Rotating the diagonal produced a 1:√2 rectangle. This method was repeated over and over to realize the other fingers, the palm, the hand, etc.

The Roman architect who incorporated Pythagorean principles directly into his designs was the first century artist and engineer Marcus Vitruvius Pollio, known commonly as Vitruvius. In his treatise, *De architectura*, translated as *The Ten Books of Architecture* (see 1914), Vitruvius describes the "perfect proportion" in architecture and sculpture as one that is based on stability, utility, and beauty. Inspired by the three Greek architectural orders—Doric, Ionic and Corinthian—Vitruvius believed that the same pattern of proportions in these orders could be applied to drawing the human body—leading to Leonardo da Vinci's Vitruvian Man (discussed below), a male nude in-scribed in the circle and the square, the two geometric archetypes of antiq-uity, connected by the squaring-the-circle method that was used to prove π, as we have seen.

A famous example of the incorporation of Pythagorean principles in archi-tecture is the Porta Maggiore Basilica, a subterranean basilica which was built during the reign of the Roman emperor Nero as a secret place of worship for neo-Pythagoreans and discovered in 1917 near Porta Maggiore in Rome (Joost-Gaugier 2006: 154–158). Each table in the sanctuary provides seats for seven people and three aisles lead to a single altar, symbolizing the three parts of the soul:

Porta Maggiore Basilica

Another example is Hadrian's Pantheon in Rome. The temple's circular structure, central axis, hemispherical dome, and alignment with the four cardinal directions symbolize the Pythagorean view of the connection among several geometric archetypes, including the sphere, the square, and the rectangle. The dome and other parts have these archetypal figures directly built into them, as can be seen below in an eighteenth-century painting of the Pantheon by Giovanni Paolo Panini:

Interior of Hadrian's Pantheon, by Giovanni Paolo Panini, Rome, c. 1734
SOURCE: WITH KIND PERMISSION BY THE NATIONAL GALLERY OF ART, WASHINGTON, DC,
SAMUEL H. KRESS COLLECTION. ACCESSION NUMBER 1939.1.24. OPEN ACCESS

In the Renaissance, Pythagoreanism was revived by several mathematicians and artists, such as Leon Battista Alberti. In his 1435 work, *De Pictura* (see Grayson 1972), Alberti suggests that the ratios by which musical sounds please our ears are the same ones by which paintings and sculptures please our eyes. Alberti used the science of classical optics as a geometric model of artistic representation, thus aligning himself with artists of his era who were developing the art of perspective drawing. A diagram in his *De Pictura* shows how to transform a circle into an ellipse, constituting a pre-topological approach to geometry:

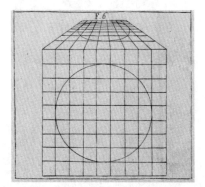

Alberti's circle-to-ellipse diagram
SOURCE: WIKIMEDIA COMMONS

Alberti was also one of the first in the Renaissance to encourage the use of the golden ratio in architecture, along with mathematician Luca Pacioli, who called it the *Divina proportione* (1509). Pacioli gave an example of how to draw the human face according to ϕ, claiming that this made a face beautiful or handsome:

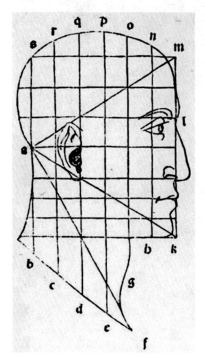

Pacioli's illustration of the human face
SOURCE: WIKIWAND

It was Raphael (Raffaello Sanzio) who utilized the golden ratio brilliantly in the same era in what is considered to be his masterpiece, *The School of Athens* (1509–1511), a fresco in the Apostolic Palace in the Vatican, in which he paints the classical scholars of Greece, from Pythagoras to Plato, bringing them together socially as if in some form of philosophical dialogue or debate:

Raphael's School of Athens (1509–1511)
SOURCE: WIKIMEDIA COMMONS

The golden ratio is found throughout the painting, including in the actual framed scene which is a golden rectangle. The various rectangular designs in the scene are also golden, as are the columns, arches, stairs, and various decorative elements.

It should be mentioned that there is a huge literature on the use of the golden ratio in art and architecture, which cannot possibly be examined here (Livio 2002). An illustrative example is the rectangular face of the front of the Parthenon in Athens (discussed briefly previously), which is purported to have a façade whose sides are in the golden ratio:

The Parthenon and the golden ratio
SOURCE: WIKIPEDIA

As in the case of the Great Pyramid at Giza, there is controversy over this architectural analysis. Nevertheless, its appearance in the design (with approximate value) seems to produce an aesthetic effect that, say, a box-like structure would not, hence indirectly suggesting that its use by the ancient builders was not casual. Kimberly Elam (2001: 6) maintains that there is a human preference for the golden ratio that comes out constantly in human aesthetic constructions, whether consciously realized or not.

Another Renaissance artist and architect who incorporated Pythagoreanism directly into his architectural philosophy and methodology was Andrea Palladio, who designed the famous Villa Capra Rotunda. Palladio promoted the Pythagorean ideals of harmonic proportions and symmetrical design in his influential 1570 treatise, *I quattro libri dell'architettura* ("The Four Books of Architecture"), in which he sets out rules that architects should follow, including what he called the set of the seven most beautiful and harmonious proportions to be used in art and architecture wherever and whenever possible. These are shown below:

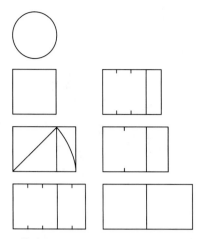

Palladio's seven proportions

As the foregoing discussion implies, Pythagoreanism has been widespread in art and architecture, starting in antiquity and culminating in the Renaissance when geometry, art, and philosophy were combined into a singular view of aesthetics. Needless to say, the use of geometric principles in architecture is not limited to western architecture. It is common across cultures and even predates Pythagoreanism itself. Domes, for instance, are found in the designs of ancient Persian and Chinese buildings, as well as in Byzantine and medieval Islamic architecture. The appeal of the dome form continues on in the contemporary world, as can be seen in the designs of legislative chambers, sports stadiums, and a variety of other buildings. This is suggestive anecdotal evidence that the sphere (based on circularity) is an archetype that has unconsciously guided architecture through the ages and, as a corollary, that Pythagoreanism is not constrained to a time and place, but is a worldview that has a basis in human cognition.

3 The Circle in Art and Symbolism

In his classic drawing of *Vitruvian Man*, Leonardo da Vinci presents us with a masterful realization of Vitruvius's principles of proportion enunciated in Book III of *De architectura*:

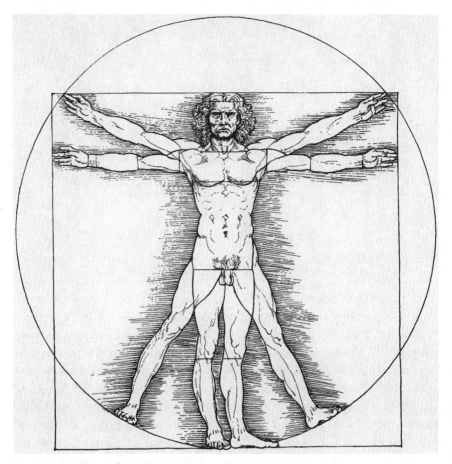

Vitruvian Man (Leonardo Da Vinci, c. 1490)
SOURCE: WIKIMEDIA COMMONS

Some of these principles can be seen in the following features of the drawing:

1. With arms outstretched, Vitruvian man is as wide as he is tall, with his genitals at the midpoint.
2. His knees are halfway between the genitals and his feet and, in symmetrical proportion, his chest is halfway between his genitals and his head.
3. His chest's width is a quarter of Vitruvian Man's height, as is the distance from his elbow to his fingertips.
4. His nose is halfway between the hairline and chin.
5. The eyebrows are halfway between the nose and hairline. The lips are halfway between the nose and chin.
6. The head is one-eighth the total height, the hand one-tenth, and the foot one-sixth.

It is significant that Da Vinci puts the ideal body that Vitruvian Man represents inside two archetypal geometric figures, the circle and the square. The use of these figures is actually intrinsic to the design of the human body itself, as indicated in another part of the *De architectura* (see Vitruvius 1914), which states that: "And just as the human body yields a circular outline, so too a square figure may be found from it. For if we measure the distance from the soles of the feet to the top of the head, and then apply that measure to the outstretched arms, the breadth will be found to be the same as the height, as in the case of plane surfaces which are perfectly square." Leonardo's collaboration with Pacioli (above), the author of *Divina proportione*, have led some scholars to suggest that he incorporated the golden ratio in drawing the proportions of Vitruvian Man (Ivins 1946), even though no such mention of the ratio is found in any of Leonardo's writings about his painting. Nevertheless, although the proportions in which the body is drawn do not match the golden ratio precisely, they come close to it, lending some substance to this hypothesis.

The circle form reverberates with symbolic meanings, of which da Vinci was certainly aware (Janson 1995). It is found across cultures with symbolic meanings. An example is the way in which Shiva Nataraja, the Hindu lord of the dance, is traditionally portrayed as dancing within a circular arch, as the following statue from eleventh century India shows (Campbell and Moyers 1988: 226):

Shiva Nataraja
SOURCE: WIKIMEDIA COMMONS

The statue represents the sacred meaning of dancing within a circle that emits flames from its circumference (*prabha mandala*), representing the cosmic fire from which existence originated and which consumes everything in the cycle of life. The source of fire are the *makara*, which are water creatures in Hindu mythology. Shiva's smile has been explained as standing for calmness in the face of contrasting energetic forces in the world.

Such representations have typically been classified in philosophy and geometry under the rubric of *sacred geometry*, which assigns mystical or symbolic meanings to geometric shapes (such as the circle) and proportions (such as the golden ratio). Sacred geometry is based on the belief that the creator of the world was a geometer—a belief that likely started with Plato. Sacred geometrical designs and constructions are evident in many religious structures such as churches, temples, and mosques. Circles are among the oldest sacred geometric symbols, generally representing unity, wholeness, and infinity. Pythagoras called the circle a "monad," the most perfect of geometric forms, without beginning or end, without sides or corners. In some traditions, standing within a circle is thought to protect people from all dangers and to keep what is inside the circle from being released.

Another famous example of the sacredness of the circle is its use in the *Mandala,* which is Sanskrit for "circle" and a key symbol in Hinduism, Buddhism, Jainism and Shintoism, representing the eternal repeating entities of reality (Brauen 1997). The basic design of most Mandalas contains a square with four gates around a circle—constituting a classic example of the blending of the square and circle archetypes. Below is an example of a Mandala in honor of the god of the cosmos, Vishnu:

Mandala of Vishnu
SOURCE: WIKIMEDIA COMMONS

The circle has also been part of western sacred architecture. One example is the South Rose Window, called the Wheel Window, of the Chartres Cathedral in France:

South Window, Chartres Cathedral
SOURCE: WIKIMEDIA COMMONS

The cathedral actually has three rose windows. The western one depicts the Last Judgment. A central oculus shows Christ as the Judge who is surrounded by a circular inner ring of 12 paired roundels containing angels and the elders of the apocalypse and an outer ring of 12 roundels showing the dead coming out from their tombs and the angels blowing trumpets to summon them to judgment. The south one above is dedicated to Christ, who is shown in the central oculus, raising his right hand in benediction, surrounded by angels. Two outer rings of twelve circles each contain the 24 elders.

An early example of how the circle is connected to mythic traditions is the ouroboros ("tail swallower"), which shows a snake feeding off its own tail. It is a fundamental symbol in alchemy. The drawing below is from a late medieval Byzantine alchemical manuscript:

Ouborous in medieval alchemy
SOURCE: WIKIMEDIA COMMONS

First attested in Egypt in the tenth century BCE, the ouroboros represents the cycle of rebirth, completion, and regeneration (Hornung 1982); it is also found in Aztec and Norse mythologies. Not surprisingly, Carl Jung saw the ouroboros as an archetype and thus part of our collective unconscious (Jung 2000, volume 14: 513):

> The alchemists, who in their own way knew more about the nature of the individuation process than we moderns do, expressed this paradox through the symbol of the Ouroboros, the snake that eats its own tail. The Ouroboros has been said to have a meaning of infinity or wholeness. In the age-old image of the Ouroboros lies the thought of devouring oneself and turning oneself into a circulatory process, for it was clear to the more astute alchemists that the *prima materia* of the art was man himself. ... which unquestionably stems from man's unconscious.

Since it is based on an archetype, it should come as little surprise that the ouroboros has been a suggestive one in science as well. The German organic chemist August Kekulé discovered the structure of benzene after seeing a drawing of the ouroboros. He writes about it as follows (from Read 1957: 179–180):

> I was sitting, writing at my text-book; but the work did not progress; my thoughts were elsewhere. I turned my chair to the fire and dozed. Again the atoms were gamboling before my eyes. This time the smaller groups kept modestly in the background. My mental eye, rendered more acute by the repeated visions of the kind, could now distinguish larger structures of manifold conformation: long rows, sometimes more closely fitted together; all twining and twisting in snake-like motion. But look! What was that? One of the snakes had seized hold of its own tail, and the form whirled mockingly before my eyes. As if by a flash of lightning I awoke; and this time also I spent the rest of the night in working out the consequences of the hypothesis.

The circle's symbolism is not limited to sacred geometry or to ancient art. A contemporary artist who has used the circle as an intrinsic part of his style is the Russian abstract painter Wassily Kandinsky. Among his most famous paintings are *Squares with Concentric Circles* (1913) and *Several Circles* (1926). The former connects the two geometric archetypes of the square and the circle and the latter depicts a square with no edges, arguably constituting a subtle reference to the squaring-the-circle technique for estimating the value of π. Below is Kandinsky's *Circles in a Circle* (1923), which includes lines that form X's, and a large X that is juxtaposed over the entire figure:

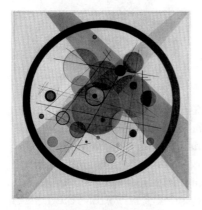

Circles in a Circle, Kandinsky 1923
SOURCE: WIKIMEDIA COMMONS

The X symbol is in opposition with the circle symbol—the latter implies continuity and infinity, whereas the former suggests contrast and transgression

(Danesi 2009, Pelkey 2017). The particular design of the X—a cross symbol that has been rotated 45 degrees—reverberates with a tension of meaning. No such tension emanates from perceiving the circular form. As Marina Roy (2000) has pointed out, the symbol X taps into a complex system of meanings that reaches back to the origins of writing. Among the first to recognize its strange appeal was Plato, who observed in the *Timaeus* that X represented the oppositional substance of the universe in which we live. On the other hand, Plato also saw the circle as a model of perfection. One could suggest that Kandinsky's painting is an indirect paean to Plato.

4   Pi in Art

Art works based on π itself are infrequent. The most common type of "Pi-Art," as it is called, is the actual representation or reproduction of the letter π, as can be seen in the mosaic outside the mathematics building at the Technische Universität Berlin:

Mosaic, Technische Universität Berlin
SOURCE: WIKIMEDIA COMMONS

This is essentially a form of public (poster) art, directing attention to the implicit importance of π itself. Another genre of Pi-Art inheres in paintings where the digits in π are drawn in some configuration. The paintings of Martin Krzywinski, who is a scientist who has contributed significantly to cancer research with his work on the graphic representation of data, are among the most well known in this genre. In one of his paintings, he portrays the digits of π across a circle with a chord. The same artistic approach to π has been taken by astronomer Nadieh Bremer, who generates "maps" of π, following up, as she indicates, on an idea put forth by mathematician John Venn in his 1888 book, *The Logic of Chance*, in which he suggests that the digits 0 to 7 as they occur in π represent eight compass directions. Bremer's art has been called "random

walk" art, an idea introduced in 1905 by Karl Pearson, and which was the subject of a famous problem by mathematician George Pólya in 1921:

> Choose a point on a graph at the beginning. What is the probability that
> a random walker will reach it eventually? Or: What is the probability that
> the walker will return to his starting point?

Pólya proved that the answer is 1, making it a virtual certainty. He called it a
1-dimensional outcome. In higher dimensions this is not the case. A random
walker on a 3-dimensional lattice, for instance, has a much lower chance of
returning to the starting point (p = 0.34). Random walk art aims to show how
randomness can be visualized. For this reason, it has had applications in science. In the diagram below the random walk technique was used to generate a
picture of Brownian Motion, the random movement of microscopic particles
in a fluid, which results from the continuous bombardment from molecules—
in the diagram the points that are more frequently traversed are darker:

Random walk for Brownian Motion
SOURCE: WIKIMEDIA COMMONS

As mentioned at the start of this chapter, π appears in the construction of the Great Pyramid at Giza, whether or not this was intentional on the part of the ancient architects. Recall that it can be detected in the ratio of the perimeter of the base to its height: 1760/280 = 2π. As historians of mathematics Roger L. Cooke (2011: 238) and Eric Temple Bell (1940: 40) maintain, it is unlikely that the builders of the Great Pyramid intentionally incorporated π in their construction projects. On the other hand, Michael Rice (2003: 24) argues that the Egyptians were well acquainted with ratios such as π and the golden ratio, as works such as the *Rhind Papyrus* indicate (Chapter 1). Whatever the case, the appearance of π in the Great Pyramid is nevertheless an interesting one, suggesting that geometric archetypes, even if coincidental or unconscious, have played a role in both art and architecture since antiquity.

5   Epilogue

Any discussion of π and the golden ratio as appearing in art and architecture raises fundamental questions that hermeneutic geometry would need to approach critically or at least carefully. One of these is the so-called "aesthetic value" of a ratio such as the golden one. The first psychological examination of this thesis is the one by Gustav Fechner (1876), who looked at the possibility that the ratio does indeed influence people's perception of something as beautiful. Fechner simply asked participants to rate how aesthetically pleasing they found a series of squares and rectangles. He found that the greatest majority selected the rectangle that approximates the golden rectangle. However, later studies testing this hypothesis have been inconclusive (see Livio 2002: 7; for an overview of the relevant studies see Iosa, Morone, and Paolucci 2018).

Recall as well from above that the Parthenon's façade is said by some to be circumscribed by a golden rectangle (Van Mersbergen 1998). However, there is also doubt about this. As mathematician Keith Devlin (2005: 54) observes: "Certainly, the oft repeated assertion that the Parthenon in Athens is based on the golden ratio is not supported by actual measurements. In fact, the entire story about the Greeks and golden ratio seems to be without foundation." Of course, there really is no way to enter into the minds of the builders of the Parthenon, nor is it really possible to assess why a golden rectangle appears to many to be more pleasing than any other kind of rectangle. Cultural factors can never be discounted. Does the same aesthetic effect occur in cultures where the golden ratio and other similar constructs are not part of their historical traditions?

We can only be sure of making claims of this kind when the artists themselves have explicitly incorporated a geometric archetype into their techniques.

Salvador Dalí, for instance, used it in his masterpiece, *The Sacrament of the Last Supper* (Livio 2002). The dimensions of the canvas are a golden rectangle. A huge dodecahedron, in perspective so that edges appear in golden ratio to one another, is suspended above and behind Jesus and dominates the composition.

Hermeneutic geometry not only investigates the presence of geometric principles in nature and the physical world, but also in art and architecture traditions. Ratios such as π and φ are connected to geometric archetypes—the circle, the spiral—which are in turn connected to others—the square, the triangle, the hexagon, etc. The fact that these are used in artistic and architectural practices, from sacred geometry to contemporary symbolic art styles, suggests that geometry is as much an aesthetic as a scientific enterprise.

Some contemporary art styles are based directly on geometry. One of these is Cubism which incorporates geometric forms directly into its techniques (Herbert 1968). The painting below by Juan Gris, titled *Le petit déjeuner*, shows how the objects in a common scene of everyday life are built with geometric ideas:

Le petit déjeuner (1914, Juan Gris)
SOURCE: WIKIMEDIA COMMONS

In effect, the Cubists were aware of the link between geometry and human affairs, recalling its use in the Renaissance to influence various artistic styles.

As John Berger (1965: 73) indicates, one cannot over-estimate the importance of Cubism to modernity:

> It is almost impossible to exaggerate the importance of Cubism. It was a revolution in the visual arts as great as that which took place in the early Renaissance. Its effects on later art, on film, and on architecture are already so numerous that we hardly notice them.

Cubists such as Pablo Picasso and Jean Metzinger were influenced by mathematician Henri Poincaré's *Science and Hypothesis* (1902), which dealt with the connections among geometry (Euclidean and non-Euclidean), physics, and nature (see Miller 2001). The idea that these connections could be expressed in visual art laid the foundations of Cubism.

CHAPTER 4

Pi in Popular Culture

> Mathematics is as much an aspect of culture as it is a collection of algorithms.
>
> CARL BOYER (1906–1976)

∴

1 Prologue

An insightful collection of essays, titled *Mathematics in Popular Culture: Essays on Appearances in Film, Fiction, Games, Television and Other Media* (Sklar and Sklar 2012), has shown that mathematics has entered the domain of contemporary popular culture in an extensive manner. Movies, books, stage plays, television programs, and websites that are either based on mathematics, or have interweaved mathematics into their narratives, are commonplace. Even a schematic list, such as the one below, will indicate how sizable the incorporation of mathematical themes, personages, etc. into pop culture has become:

TABLE 1　Mathematics in popular media

Movies	Books	TV/Internet
Straw Dogs (1971)	*Ratner's Star* (1992)	Various episodes of *The Simpsons*
It's My Turn (1980)	*Picasso* (1993)	*Math Country* (1970s)
Sneakers (1992)	*Arcadia* (1993)	*Numbertime* (1993)
Antonia's Line (1996)	*Copenhagen* (1999)	*In Her Own Words* (1991)
Good Will Hunting (1997)	*Hypatia* (2000)	*Solving Fermat* (1997)
Pi: Faith in Chaos (1998)	*Proof* (2000)	*MIT's Tech Talk* (1998)

© KONINKLIJKE BRILL NV, LEIDEN, 2021 | DOI:10.1163/9789004433397_005

TABLE 1   Mathematics in popular media (*cont.*)

Movies	Books	TV/Internet
Enigma (2001)	*QED* (2001)	*Futurama* (1999–2008)
A Beautiful Mind (2001)	*Fermat's Last Tango* (2001)	*John Nash* (2002)
Enigma (2001)	*Uncle Petros and Goldbach's Conjecture* (2001)	*Numb3rs* (2005–2010)
Fermat's Room (2007)	*The Fractal Murders* (2002)	Various episodes of *The Big Bang Theory* (2007–2019)
Flatland (2007)	*The Curious Incident of the Dog in the Night-time* (2003)	*The Story about Maths* (2008)
The Calculus of Love (2011)	*Victoria Martin: Math Team Queen* (2007)	*Numberjacks* (2009)
The Traveling Salesman (2012)	*Pythagoras' Revenge: A Mathematical Mystery* (2009)	*The Code* (2011)
The Theory of Everything (2014)	*The Housekeeper and the Professor* (2009)	*Numberphile* (2011)
The Imitation Game (2014)	*The Devotion of Suspect X* (2011)	*Hard Nut* (2013)
The Man Who Knew Infinity (2015)	*Lost in Math: How Beauty Leads Physics Astray* (2018)	*Odd Squad* (2014)
Hidden Figures (2016)	*The Joy of X: A Guided Tour of Math from One to Infinity* (2019)	*The Great Math Mystery* (2015)
Gifted (2017)	*Humble Pi: When Math Goes Wrong in the Real World* (2020)	Hundreds of websites dealing with math for a general audience: see www.youtube.com/watch?v=yRlsmiECG4U

Of specific relevance to the present discussion is the 1998 film π: *Faith in Chaos*, where π has the "lead role." Given its significance, the discussion will be postponed to the next chapter. The purpose of this one is threefold: (1) to take a look at how mathematics has become inserted into pop culture and why this is so; (2) to discuss the current fascination with π, as can be seen with the advent and celebration of Pi Day; and (3) to consider the reasons for the continuing obsession with calculating π to larger and larger digits with computer algorithms.

2   Mathematics in Popular Culture

Before dealing with π in popular culture, it is useful to take a generic look at the plausible reasons why mathematics itself has become a fixed part of popular culture in all media, from print to film and the Internet. The "math movie," for instance, has evolved, since early cinema, into a genre all its own, shattering the implicit social myth that math is too intellectual and thus uninteresting for a mass audience (Polster and Ross 2012). This genre can itself be divided into two main subgenres: (a) films that revolve around a math theme or personage, and (b) films that incorporate mathematics at some point in the plot. Among the numerous (a)-genre movies, the following three can be mentioned here as exemplary cases-in-point: *A Beautiful Mind* (2001), *Proof* (2005), and *The Imitation Game* (2014). All three are united by a common subtext—math geniuses are eccentric and may suffer from mental issues as a result.

A Beautiful Mind is a narrative portrait of American mathematician John Forbes Nash (1928-2015). There is actually little mathematics in the movie. As other (a)-genre movies, it focuses primarily on a prominent math personage. It is based in part on Sylvia Nasar's (1998) biography of Nash, who made an astonishing breakthrough in game theory early in his career—the development of mathematical models of strategic interaction among decision-makers (Myerson 1991)—for which he won the Nobel Prize in 1994 in economics. The film's overall theme is that the effort and obsessive concentration to arrive at his discovery may have led to Nash's mental health problems, which included delusions and paranoia, starting in 1959. Remarkably, by the time he won the Nobel, Nash had recovered from his mental problems, regaining the ability to function socially and emotionally by, ironically, dedicating himself even more intensively to mathematics, perhaps feeling comfortable in the world of numbers and their ability to take a "beautiful mind," such as his, away from the stress of daily emotional life into a (Pythagorean) realm of metaphysical exploration. But ultimately it is not mathematics, but human love that conquers everything, as Nash himself observes in his Nobel Prize speech in the movie, in which he alludes to the love of his wife:

> What truly is logic? Who decides reason? My quest has taken me to the physical, the metaphysical, the delusional, and back. I have made the most important discovery of my career—the most important discovery

of my life. It is only in the mysterious equations of love that any logic or reasons can be found. I am only here tonight because of you. You are the only reason I am. You are all my reasons. Thank you.

Nash also is aware of the role of "madness" in his accomplishments: "Without his 'madness,' Zarathustra would necessarily have been only another of the millions or billions of human individuals who have lived and then been forgotten."

Proof also links eccentricity, mental states, and mathematics. The daughter of a famous, but troubled, mathematician suspects that she has inherited his mental problems. The plot is replete with episodes where mathematical theorems are used and proof is dissected in easy-to-understand ways, making the movie more than just a biographical one, stressing the importance of mathematical thinking to human life. The movie is a version of David Auburn's play, *Proof* (2000), about a genius mathematician, Robert Llewellyn, who had passed away just before the film begins. Llewellyn had made significant contributions to several areas of mathematics in his early twenties. But these came at an emotional price, since they made him mentally unstable until his death at the age of 63. Llewellyn's daughter, Catherine, had sacrificed her own career as a mathematician to take care of her father. After his death, she starts to experience anxieties that lead her to question her own sanity, believing that she may have inherited her father's neuroses. Her older sister, Claire, who had carved out a successful career as a Wall Street analyst, wants Catherine to sell her house and go back with her to New York for therapy.

In this emotionally-charged atmosphere, a young graduate student named Hal, who was Robert's student, wants to get his hands on Llewellyn's notes, to gauge if there is anything interesting in them mathematically for his own career motives. With Claire's help, Hal discovers a notebook that contains new mathematical proofs, but it is not certain if these are the ideas of Robert Llewellyn or Catherine. The fact that the authorship of the notebook is left uncertain alludes to an implicit theme in the history of mathematics, starting with the Pythagoreans—what counts it the mathematics itself, not the individual mathematician—unlike what Hal is seeking to do. Hal doubts that Catherine was the inventor of the new proof, given its difficulty and complexity, as encapsulated in the following stretch of dialogue in which it is discussed; but Catherine pushes back, admonishing Hal for the hypocritical academic game of one-upmanship that he is playing:

[HAL]	It's too advanced. I don't even understand most of it.
[CATHARINE]	You think it's too advanced?
[HAL]	Yes.
[CATHARINE]	It's too advanced for you.

[HAL]	You could not have done this work.
[CATHARINE]	But what if I did?
[HAL]	Well, what if?
[CATHARINE]	It would be a real disaster for you. Wouldn't it? You and the other geeks who barely finished their PhD's, who are marking time doing lame research, bragging about the conferences they go to. Wow. Playing in an awful band and whining that they're intellectually past it at 26, because they are.

The *Imitation Game* is about the famous mathematician, cryptographer, and computer scientist, Alan Turing, and his well-known inner struggles. Based on Andrew Hodges's biography, *Alan Turing: The Enigma* (1983), the movie, like the other two above, revolves around the mathematician-as-genius theme, vindicating the efforts of such eccentric individuals. As Pickover (2005: 14) remarks, there is a deeply-held mythology in popular culture that "mathematicians throughout history have had a trace of madness or have been eccentric," citing British mathematician John Edensor Littlewood, who suffered from depression: "Mathematics is a dangerous profession; an appreciable proportion of us goes mad." Turing died young, in part, because he was persecuted for his sexual orientation in 1952. The *Imitation Game* instantly generated broad interest in Turing's accomplishments as a mathematician and computer science pioneer. In 1936, he developed a hypothetical computing machine—now called the Turing machine—that became a model for determining what tasks a computer could (or could not) perform, thus establishing the central principles behind modern-day computers. In one episode of the movie, Turing makes the claim that thinking is not unique to humanity—an idea that is spreading broadly in both science and popular culture:

> Of course machines can't think as people do. A machine is different from a person. Hence, they think differently. The interesting question is, just because something, uh.. thinks differently from you, does that mean it's not thinking? Well, we allow for humans to have such divergences from one another. You like strawberries, I hate ice-skating, you cry at sad films, I am allergic to pollen. What is the point of ... of ... different tastes, different ... preferences, if not, to say that our brains work differently, that we think differently? And if we can say that about one another, then why can't we say the same thing for brains ... built of copper and wire, steel?

The movie also brilliantly challenges the idea of normalcy and norms—an important one especially for Turing who was persecuted for the fact that he was different sexually. In fact, a theme of the movie, which is implicit in the other two above as well, is that it is often those who are marginalized or ostracized by normal people who end up changing the world for the better.

Math movies that fall in the (b)-genre category (above)—movies in which math is mentioned or utilized at some point in the narrative—have been a part of cinema culture since the early part of the twentieth century. An example is the fantasy musical, the *Wizard of Oz* (1939), based on L. Frank Baum's novel, *The Wonderful Wizard of Oz*. The Wizard bestows the degree of "Doctor of Thinkology" upon the simple-minded Scarecrow, providing an ironic commentary on academia:

[WIZARD]   Why, anybody can have a brain. That's a very mediocre commodity. Every pusillanimous creature that crawls on the Earth or slinks through slimy seas has a brain. Back where I come from, we have universities, seats of great learning, where men go to become great thinkers. And when they come out, they think deep thoughts and with no more brains than you have! But they have one thing you haven't got—a diploma. Therefore, by virtue of the authority vested in me by the *Universitatus Committeatum E Pluribus Unum*, I hereby confer upon you the honorary degree of Th.D.

[SCARECROW]   Th.D?

[WIZARD]   That's Doctor of Thinkology.

After receiving the degree, the Scarecrow immediately recites a mathematical theorem to show off his new mental prowess: "The sum of the square roots of any two sides of an isosceles triangle is equal to the square root of the remaining side. Oh joy, rapture! I've got a brain!" The irony of the scene is palpable, constituting a critique of universities as diploma mills, rather than as places where real learning occurs. But so too is the nature of the Scarecrow's response, which makes no geometric sense.

In the movie, *In the Navy* (1941), comedians Bud Abbott and Lou Costello argue about how to make seven batches of thirteen donuts each for the Naval officers in a hilarious episode. The problem is discussed in bogus mathematical terms, perhaps as an aside to the main comedic plot or perhaps to emphasize that mathematics and humor may have been unnecessarily disconnected in the popular imagination. The simple answer is rendered as a comedic routine (from Polster and Ross 2012):

[ABBOTT]   Hey, doughnuts!

[COSTELLO]   No, Smokey! Don't!

[ABBOTT]   Oh, come on. One doughnut.

[COSTELLO]   I haven't got enough. I can't afford it. I just baked 28 of these things. Well, after all there are 7 officers I've got to feed and I've just got enough to give them 13 apiece.

Abbott asks Costello to prove to him that 7 times thirteen is twenty-eight. Here are his three hilarious proofs:

First Proof

[COSTELLO]   There were 7 officers. There's a 7. Now I'm going to divide to prove it to you. Now, 28 doughnuts (writes 7 and 28 on a blackboard). Now, 7 into 2. You couldn't even push that big 7 into that little 2. Therefore we can't use the 2. I'm gonna let Dizzy hold it ... I'll use it later. Now, 7 into 8 (writes down a 1) ... Now, we're gonna carry the 7. It's getting a little heavy, so I'll put it right down there: 7 from eight ... 1. Now, a minute ago, we didn't use the 2. I'm gonna use it now. Dizzy, give me back the 2. Thanks. Put it right down there. Now 7 into 21?

[ABBOTT]   Three times.

[COSTELLO]   7 ... 28 ... 13.

[ABBOTT]   Now wait a minute!

Second Proof

[ABBOTT]   Put down 13 up there. Now you claim that each officer gets 13 doughnuts? Put down 7, draw a line. Now 7 times 13 is what?

[COSTELLO]   28.

[ABBOTT]   Prove it.

[COSTELLO]   7 times 3?

[ABBOTT]   21.

[COSTELLO]   7 times 1

[ABBOTT]   7

[COSTELLO]   7 and 1?

[ABBOTT]   1.

[COSTELLO]   Two.

[ABBOTT]   Two. Oh no. Come on.

Third Proof

[ABBOTT]   We add this up. Put down 13 seven times. Now we're getting it. You claim all this added up amounts to what?

[COSTELLO]   28.

[ABBOTT]   Adding up the 3's from the bottom to the top: 3, 6, 9, 12, 15, 18, 21.
[COSTELLO]   Just keep adding the 7 ones from the top to the bottom: 22, 23,
                24, 25, 26, 27, 28!

The convoluted proofs are humorous, not only because they play with numbers incorrectly, but also because they ultimately imply that proof is a human
construct, and can thus be manipulated by humans any way they want.

The romantic comedy, *It's My Turn* (1980), starts with the protagonist, mathematics professor Kate Gunzinger, giving a lesson on the snake lemma of ho-
mological algebra to an insufferable graduate student (Wilson 2005). Math
themes come up in a few other parts of the movie. For instance, the concept of
prime number comes up after Gunzinger runs into the son of one of the characters in the movie named Homer. Here is the relevant dialogue (from Polster
and Ross 2012):

[KID]   A prime number is one that cannot be expressed as a product
                of two smaller numbers.
[KATE]   Is 2 a prime number?
[KID]   Yes.
[KATE]   Is 3?
[KID]   Yes.
[KATE]   Is 4?
[KID]   No.
[KATE]   Why?
[KID]   2 times 2.
[HOMER]   Smart? Smart kid. You are a graduate of Harvard?
[KID]   No, I went to Yale.

The dialogue unfolds like a Socratic one, where questions are designed to elic-
it answers that lead to self-knowledge. This type of dialectic reasoning is the
basis of mathematical and philosophical inquiry, captured in this dialogical
snapshot by the movie.

In *Stand and Deliver* (1988), teacher Jaime Escalante (a real math educator)
believes that math is the means through which his students can escape from
the poverty and degradation of the Barrio. The students show determination in
taking the Advanced Place Calculus test, doing so well that the College Board
thinks that they must have cheated. The subtext is conspicuous—gaining skill
at mathematics is a step in the direction away from poverty towards success
and discriminatory attitudes as well.

Like the movies, TV has also incorporated mathematics in individual programs and series. "TV math," as it can be called, falls into several areas. One of these is documentary programs revolving around some mathematical idea or some mathematician, which are therefore like the (a)-genre of movie math, and thus requires little commentary here. The (b)-genre type of movie finds many counterparts in TV math. An example is *Futurama*, which was an animated series that aired from 1999 to 2003. It was revived for 26 new episodes between 2010 and 2011 and then again for a second set between 2012 and 2013. Although it did not focus exclusively on math, it brought math into episodes regularly. Several members of the writing team for *Futurama* were actual mathematicians.

Another example is *The Simpsons,* as Simon Singh has demonstrated in his book, *The Simpsons and Their Mathematical Secrets* (2013). Given the humorous context in which math occurs in this sitcom viewers are made to feel comfortable about ideas that might otherwise seem complex. Three examples provided by Singh are the following:

1. In "Homer Cubed" which is part of "Treehouse of Horror VI" (1995), the unsolved problem of determining which problems are hard and which are easy, known as P = NP, in computer science, is discussed at some length, as is the hexadecimal system of numbers—numbers to the base 16 which are used in ASCII (American Standard Code for Information Interchange).

2. Fermat's Last Theorem is discussed in the episode titled "The Wizard of Evergreen Terrace" (1998), in which Homer seems to write a simple solution on a blackboard, which is, however, just slightly off.

3. Prime numbers are featured in the episode called "Marge and Homer Turn a Couple Play" (2006), in which we see the prime number 8,191 on a large screen at Springfield stadium—the number is a so-called Mersenne prime number.

Another TV math genre, exemplified by the program *Numb3rs* (2005 to 2011), can be called "forensic math," since it involves the use of math in solving crimes. The program followed the exploits of an FBI agent, Don Eppes, and his brother, Charlie, who helped Don solve crimes with the use of mathematics. An episode started typically with a crime to be investigated by Don who then seeks Charlie's help. The latter develops a mathematical model that pertains to the case. Actual mathematicians worked as consultants and real math was used in the program. In *The Numbers Behind Numb3rs: Solving Crime with Mathematics* (2007), Keith Devlin and Gary Lorden (the main consultant used for the show) demonstrate how the math techniques on the program were based on actual uses by law enforcement agencies to catch criminals.

A third subgenre consisting of books on mathematics for a broad audience can be called simply "pop math." The books written by the late Isaac Asimov starting in the 1950s are primary examples. These captured a wide audience for mathematics. The titles of Asimov's books themselves provide a hermeneutic commentary on the kind of themes that characterize pop math: *Realm of Numbers* (1959), *Realm of Measure* (1960), *Realm of Algebra* (1961), *Quick and Easy Math* (1964), *An Easy Introduction to the Slide Rule* (1965), and *How Did We Find Out About Numbers?* (1973)

As Frank Nuessel (2013) has argued, pop math constitutes a form of literature that aims to show the importance of math in an entertaining fashion. He identifies four main subgenres:

1. works presenting math in an easy-to-read fashion
2. biographical novels about mathematicians
3. collections of math puzzles for general entertainment
4. fictional novels based on math

A classic book in category-(1) is the 1940 one by James Kasner and John Newman, *Mathematics and the Imagination*. The authors show, in non-technical language, how mathematics is the result of blending logic and imagination. The book is also famous for having introduced the terms *googol* and *googolplex*. The former is a number written as 1 followed by 100 zeros, or 10^{100}. A *googolplex* is an even larger number, defined as 10 to the googol power, or 10 multiplied by itself a googol number of times:

$$10^{googol} = 10^{10^{100}}$$

It is mind-boggling to envision the number this represents, let alone write it out even in part. It has, in fact, been estimated that there are more zeroes in a googolplex than there are particles in the known universe. Interestingly, the term *googol* was the source of the name for Google, constituting an accidental misspelling of *googol*. The name was selected purportedly because the search engine is intended to provide enormous quantities of information. Another classic book in category-(1) is the one by the late Reuben Hersh, *What Is Mathematics, Really?* (1998). Hersh portrays math as a social phenomenon, tied to historical forces, not as something intellectually abstract that is not grounded in any cultural paradigm.

Books in category-(2) above are about the lives and achievements of famous mathematicians—corresponding to the (a)-genre movies. The 1992 book by Robert Kanigel, *The Man Who Knew Infinity: A Life of the Genius Ramanujan*, recounts the amazing story of an unschooled Indian clerk, Srinivasa Ramanujan, who wrote a famous letter in 1913 to mathematician G. H. Hardy discussing his ideas about number theory. From this, a collaboration started between the two, leading to many new explorations in advanced mathematics. The narrative has a tragic nuance to it, since, according to the author, Ramanujan's creative intensity eventually took its toll, leading to his death at the age of thirty-two, leaving behind a legacy that is still being plumbed for its many insights to this day.

Category-(3)—also known as recreational or puzzle mathematics—consists of collections of ingenious puzzles, from classic puzzle makers to recent makers of puzzles. Although the use of puzzles in math goes right back to antiquity, their migration to pop culture is a modern-day phenomenon. Like other genres of pop math, the subtext in these books is that math is both enjoyable and revelatory of how the mind works. For example, in their interesting book, *Taking Sudoku Seriously* (2011), Jason Rosenhouse and Laura Taalman show how the inner logic of Sudoku puzzles mirrors how mathematicians think in schematic form

Category-(4)—fictional or semi-fictional narratives that deal with mathematical topics (Mann 2010, Sklar and Sklar 2012)—may actually be traced back to Lewis Carroll's *Alice's Adventures in Wonderland* (1885) and *Through the Looking-Glass, and What Alice Found There* (1871), which are replete with mathematical ideas (Devlin 2010). Carroll alludes to the concept of quaternions, which are complex numbers made up of real and imaginary numbers that satisfy certain conditions. These were discovered before the publication of Carroll's books by Irish mathematician William Rowan Hamilton. As Bayley (2009) notes, the parallels between Hamilton's numbers and the Mad Hatter's tea party are striking. In the scene, Alice is seated at a table with three characters—the Hatter, the March Hare and the Dormouse. The character Time, who has had a falling out with the Hatter, is absent. Bayley interprets the scene as follows: "the members of the Hatter's tea party represent three terms of a quaternion, in which the all-important fourth term, time, is missing. Without Time, we are told, the characters are stuck at the tea table, constantly moving round to find clean cups and saucers."

Another book in category-(4), published at about the same time, is the novella, *Flatland: A Romance of Many Dimensions* (1884). It was written by the preacher and literary critic Edwin A. Abbott. The characters of the novel are geometrical figures living in a two-dimensional universe called Flatland. They see each other edge-on, and thus as dots or lines, even though, from the vantage point of an observer in three-dimensional space looking down upon them from above, they are actually lines, circles, squares, triangles, etc.

In his first chapter of Part I ("This World"), Abbott (1884: 34–35) provides a description and an accompanying graphic to describe Flatland. It is worth repeating in its entirety here since it provides a snapshot of the novel and its hermeneutic geometrical purpose:

> I call our world Flatland, not because we call it so, but to make its nature clearer to you, my happy readers, who are privileged to live in Space.
>
> Imagine a vast sheet of paper on which straight Lines, Triangles, Squares, Pentagons, Hexagons, and other figures, instead of remaining fixed in their places, move freely about, on or in the surface, but without the power of rising above or sinking below it, very much like shadows— only hard and with luminous edges—and you will then have a pretty correct notion of my country and countrymen. Alas, a few years ago, I should have said "my universe": but now my mind has been opened to higher views of things.
>
> In such a country, you will perceive at once that it is impossible that there should be anything of what you call a "solid" kind; but I dare say you will suppose that we could at least distinguish by sight the Triangles, Squares, and other figures, moving about as I have described them. On the contrary, we could see nothing of the kind, not at least so as to distinguish one figure from another. Nothing was visible, nor could be visible, to us, except Straight Lines; and the necessity of this I will speedily demonstrate.
>
> Place a penny on the middle of one of your tables in Space; and leaning over it, look down upon it. It will appear as a circle.
>
> But now, drawing back to the edge of the table, gradually lower your eye (thus bringing yourself more and more into the condition of the inhabitants of Flatland), and you will find the penny becoming more and more oval to your view; and at last when you have placed your eye exactly on the edge of the table (so that you are, as it were, actually a Flatlander) the penny will then have ceased to appear oval at all, and will have become, so far as you can see, a straight line.
>
> The same thing would happen if you were to treat in the same way a Triangle, or Square, or any other figure cut out of pasteboard. As soon as you look at it with your eye on the edge on the table, you will find that it ceases to appear to you a figure, and that it becomes in appearance a straight line. Take for example an equilateral Triangle—who represents with us a Tradesman of the respectable class. Fig. 1 represents the Tradesman as you would see him while you were bending over him from above; figs. 2 and 3 represent the Tradesman, as you would see him if your eye were close to the level, or all but on the level of the table; and if your eye were quite on the level of the table (and that is how we see him in Flatland) you would see nothing but a straight line.

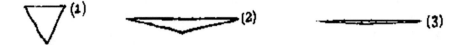

The novel has become so popular that it has generated offshoots and extrapolations, including pop math books such as *The Dot and the Line: A Romance in Lower Mathematics* (Juster 1963), *Sphereland: A Fantasy about Curved Spaces and an Expanded Universe* (Burger 1965), *The Planiverse: Computer Contact with a Two-Dimensional World* (Dewdney 1984), *Flatterland: Like Flatland, Only More So* (Stewart 2001), and *Spaceland: A Novel of the Fourth Dimension* (Rucker 2002). Most of these update the geometry that was known to Abbott during his times, including notions such as fractal geometry, topology, and hyperbolic geometry.

There are also various film versions of *Flatland*, including *Flatland* (1965), the short film *Flatland* (1982) directed by mathematician Michele Emmer, and *Flatland: The Movie* (2007), an animated film followed by *Flatland 2: Sphereland* (2012). Various television programs have alluded to the novel, including *The Outer Limits* (October 3, 1964) and *The Big Bang Theory* (January 11, 2010). Two video games have been based directly on the novel—*The Flatland Role Playing Game* and *Miegakure*.

3 Pi in Popular Culture

Pi appears in different pop culture media, such as in books that fall into Nuessel's category-(1), including Petr Beckmann's *A History of π* (1971), David Blatner's *The Joy of Pi* (1997), and Alfred Posamentier's *Pi: A Biography of the World's Most Mysterious Number* (2004). These present the mathematical and scientific facts about π in an easy-to-follow style, perhaps enhancing the mystery that π evokes in the popular imagination.

One of the more significant appearances of π in Nuessel's category-(4) is in Carl Sagan's novel *Contact* (1985), on which the movie of the same name (1997) is based. The novel recounts the fictional story of a radio astronomer, Ellie, who is contacted by aliens from the star Vega, via a message which claims that there is definitely a pattern within π, even though they have not deciphered it themselves. Acting upon a suggestion by the aliens, Ellie sets her computer program to examine the expansions of π in different number bases. She discovers a point in a string late in the expansion of π in base 11, where the digits stop varying randomly and start producing 1's and 0's, which, when properly aligned, produce a circular pattern. The aliens suggest that this pattern is built into the universe itself, yet they have not unraveled its meaning. However, it could still be a statistical anomaly. The novel and movie recall the 1952 movie, *Red Planet Mars,* in which a similar theme can be found when a young boy comes up with the idea that the decimal expansion of π could be the language that would allow us to communicate with extraterrestrials.

The relevant section of *Contact* is worth reproducing here. It involves Elaine's interaction with the aliens, who answer her questions in a way that evokes the Pythagorean sense that mathematics is the language of the universe:

[ELAINE]   I want to know about your myths, your religions. What fills you with awe? Or are those who make the numinous unable to feel it?

[ALIEN]   Certainly we feel it. I don't say this is it exactly, but it'll give you a flavor of our numinous. It concerns pi, the ratio of the circumference of a circle to its diameter. Our mathematicians have made an effort to calculate it out to ... let's say the ten-billionth place. You won't be surprised to hear that other mathematicians have gone further. Well, eventually—let's say it's in the ten-to-the-twentieth-power place—something happens. The randomly varying digits disappear, and for an unbelievably long time there's nothing but ones and zeros.

[ELAINE]   And the zeros and ones finally stop? You get back to a random sequence of digits? And the number of zeros and ones? Is it a product of prime numbers?

[ALIEN]   Yes, eleven of them.

[ELAINE]   You're telling me there's a message in eleven dimensions hidden deep inside the number pi? Someone in the universe communicates by mathematics? Mathematics isn't arbitrary. I mean pi has to have the same value everywhere. How can you hide a message inside pi? It's built into the fabric of the universe.

[ALIEN]   Exactly.

As this passage encapsulates, *Contact* intertwines mathematics, science, and philosophy in a Pythagorean way, evoking the theory of *musica universalis,* or more precisely in the book, a *lingua universalis.* Sagan suggests that we should look beyond speculation (the numinous) and look for objective evidence (mathematics) upon which to build our understanding of the universe.

Pi appears in various other fictional texts, often making brief appearances. For example, in the novel *Life of Pi* (2001) by Yann Martel, the main character, whose full name is Piscine Molitor Patel, is bullied at school over his name "Piscine" (swimming pool in French), which can be corrupted into an obscenity.

Fed up with the harassment, Patel abbreviates his name to "Pi." On the first day of school, and just before the first teacher calls out his name, he goes to the blackboard and writes it as "Pi Patel," adding "π = 3.14" and drawing a circle. He repeats this in every classroom, uttering at one point, "And so, in that Greek letter that looks like a shack with a corrugated tin roof, in that elusive, irrational number with which scientists try to understand the universe, I found refuge" (Martel 2001: 23–24). Patel adopts a name that is based in mathematics, in an effort to combat human cruelty, in the hope that the pursuit of truth, as reflected by mathematics itself, will render human malice trivial and ineffectual.

Below is a selective and schematic listing of other appearances of π in pop culture texts:

- In Alfred Hitchcock's *Torn Curtain* (1966), π is used as a secret code in Cold War political intrigue. On his way to a conference, physicist Michael Armstrong, receives a message to pick up a book in Copenhagen. The book contains a cryptic message: "Contact π in case of emergency." Armstrong had made preparations to return to the West via an escape network, known as π. Eventually, an East German security officer realized what π meant and that Armstrong was a double agent. It can be argued, albeit speculatively, that the use of π as part of a secret political network alludes, subtextually, to the secrets that π itself may enfold, but which have not yet been decoded.

- In the movie *The Net* (1995), Angela Bennett is a computer programmer who is approached by a colleague for help with a glitch he found in a client website. Bennet goes to the website, clicking on the π symbol she sees there. As a result, she is transported to the secret databases of different government agencies.

- In an episode of the Batman television series, "Penguin Sets a Trend" (1967), Batman employs the cube root of π (1.46) to calculate where he and Robin will land before being catapulted by a villain.

- In the "The Wolf in the Fold" (1967) episode of the original Star Trek TV series (mentioned in Chapter 1 as well), Mr. Spock orders a computer gone amok to compute π to the last digit, knowing full well that the computer will never stop and thus become ineffectual because, as he says, "the value of pi is a transcendental figure without resolution."

- In a *Doctor Who* TV episode, "The Five Doctors" (1983), the First Doctor employs π to help guide him across a deadly floor designed as a chessboard in the Dark Tower of Rassilon, because of a statement made by The Master that the chessboard is "easy as Pi."

- In the novel *Eon* (1985) by Greg Bear, the protagonists measure the amount of space curvature using a device that computes π, knowing that only in two-dimensional space-time will a circle have a circumference to diameter ratio of 3.14159.

- In an episode of the TV series, *Northern Exposure*, titled "Nothing's Perfect" (1992), a mathematician becomes obsessed with searching for a pattern in the digits of π, even knowing full well that it is a transcendental number.
- In the novel *Time's Eye* (2003), by Arthur C. Clarke and Stephen Baxter, a spherical device has a circumference to diameter ratio that appears to be an exact integer, 3, across all its planes, recalling ancient estimates of π.
- In a 2004 episode of the *Jimmy Neutron* TV series, "Return of the Nanobots," the main character destroys the nanobots, which were supposed to fix all errors, but were instead destroying humanity because of their relentless search for an error-free world. Victory over the nanobots is accomplished by asking them to correct a test paper in which π is given erroneously as equaling exactly 3, a task that, like the Star Trek episode, will keep them engaged in it indefinitely.
- In the novel *Going Postal* (2004), by Terry Pratchett, an inventor named Bloody Stupid Johnson invents a mail sorter with a wheel for which π is exactly 3, which is, of course, mathematically impossible. This leads to the failure of his post office and the universe at the same time, suggesting that if π were different, so would the present world, which, in effect, would not exist as such.
- In the TV show, *High School Musical* (2006), a teacher is seen writing up two of Ramanujan's formulas for $1/\pi$, whereby a student immediately spots an error in the second equation—the teacher had written $8/\pi$ instead of $16/\pi$.
- References to π are found in various episodes of *The Simpsons*. In one episode, "Lisa's Sax" (1997), two schoolmates recite the following: "Cross my heart and hope to die, here's the digits that make pi, 3.141592653589793238384." In another episode, a sign at the Springfield graveyard says, "Come for the funeral, stay for the π."

The incorporation of π in films and novels as part of themes involving mystery, intrigue, and even destiny is a contemporary engagement in unconscious Pythagoreanism. Pythagoras can retroactively be described as an ancient mythical celebrity, since he himself claimed, according to legend, the he was the son of Hermes. In another myth, Pythagoras had a golden thigh, which he would purportedly show to doubters of his divinity, thus converting them to the belief that he was divine. It is also claimed that he got people to believe that they could achieve whatever they wanted by just thinking about it—a myth that has been recycled by the bestseller, *The Secret* (2006), by Rhonda Byrne, in which she makes this specious claim (see also Martínez 2012).

4   Pi-Mania

The modern fascination with π has, in some ways, become an obsession. With a bit of poetic license, this can be called *Pi-Mania*—the use of "mania" is not intended in any way to be literal; it is simply a convenient metaphor. One symptom of Pi-Mania is the desire by some to memorize the digits of π to higher and higher levels.

Actually, this obsession reaches back considerably in time. For instance, in a circular room in the Parisian science museum, *Palais de la Découverte*, called the "Pi room," 707 digits of π are inscribed on its dome-like ceiling. The digits were taken from British school teacher William Shank's 1883 calculation of π, which, however, was only correct up to the first 527 places—an error discovered in 1944 by another English mathematician, D. F. Ferguson, and corrected a few years later (Posamentier 2004: 118):

Pi Room (Palais de la Découverte)
SOURCE: STOCK PHOTO (PALAIS)

Shank owned a boarding school and, apparently, he would spend his free time calculating π to larger and larger values, using Machin's formula:

$$\pi/4 = 4 \tan^{-1}(^1/_5) - \tan^{-1}(^1/_{239})$$

Feats of calculating and memorizing the digits of π are now classified under *piphilology* (Arndt and Haenel 2006: 44–45), defined as the creation and use of techniques that will facilitate memorizing any span of digits of π. The term is a play on the word *pi* itself and of the field of *philology* (the study of the history of languages). Some mnemonic efforts border on the unbelievable. Here are a few examples. In 1981, an Indian man named Rajan Mahadevan accurately recited 31,811 digits of pi from memory. In 1989, Hideaki Tomoyori of Japan recited 40,000 digits. And in 2005, Lu Chao of China recited 67,890 digits.

Pi-Mania may have actually started with German-Dutch mathematician Ludolph van Ceulen who had published a 20-decimal value in his 1596 book *Vanden Circkel* ("On the Circle"), which he later expanded to 35 decimals, claiming that he could recite these from memory at will. He was so proud of his accomplishment that he had the 35 digits inscribed on his tombstone.

Piphilology also covers mnemonic techniques that are devised to help recalling the digits of π. The following one was published in a 1914 edition of *Scientific American*:

See, I have a rhyme assisting my feeble brain, its tasks ofttimes resisting.

Replacing each word by the number of letters it contains yields π to 12 decimal places:

See	= 3
I	= 1
have	= 4
a	= 1
rhyme	= 5
assisting	= 9
my	= 2
feeble	= 6
brain	= 5
its	= 3
tasks	= 5
ofttimes	= 8
resisting	= 9

3.141592653589

An early example of this type of technique is the one devised by English scientist James Jeans (Arndt and Haenel 2006: 44–45):

How I want a drink, alcoholic of course, after the heavy lectures involving quantum mechanics.

Other well-known examples are (Arndt and Haenel 2006: 44–45):

> How I wish I could recollect pi easily today!
> May I have a large container of coffee, cream and sugar?
> The point I said a blind Bulgarian in France would know (the word point represents the actual decimal point in π)

This type of mnemonic technique is now called a *piem*—a blend of *pi* and *poem*. In longer examples, 10-letter words are used to stand for the digit zero, which is also applied to handle repeated digits—a method called *pilish* (*pi-lish*) writing—a type of writing in which the lengths of consecutive words match the digits of the number. The 1996 short story "Cadaeic Cadenza" by Michael Keith, in which each word in sequence has the same number of letters as the corresponding digits in π, is an example of pilish (Bellos 2015). For example, the word *cadae* is the equivalent of 3.1415 (that is, c = 3rd letter of the alphabet, a = 1st letter, d = 4th letter, e = 5th letter). Keith also composed a 10,000-word work titled, *Not a Wake* (2010), which is a collection of poems, short stories, crossword puzzles, and other texts constructed in pilish—the number of letters in successive words of the text reveal successive digits of π.

In his classic work, *Language on Vacation* (1965), Dmitri Borgmann constructed the following 30-word piem:

> Now, a moon, a lover refulgent in flight,
> Sails the black silence's loneliest ellipse.
> Computers use pi, the constant, when polite,
> Or gentle data for sad tracking aid at eclipse.

Expert memorizers typically use a strategy known as the "method of loci," which involves spatial visualization to remember information—a method that goes back to antiquity, adopted by Cicero and Quintilian among others. The method is described by John O'Keefe and Lynn Nadel (1978) as follows:

> [In] the method of loci, an imaginal technique known to the ancient Greeks and Romans ... the subject memorizes the layout of some building, or the arrangement of shops on a street, or any geographical entity which is composed of a number of discrete loci. When desiring to remember a set of items the subject 'walks' through these loci in their imagination and commits an item to each one by forming an image between the item and any feature of that locus. Retrieval of items is

achieved by 'walking' through the loci, allowing the latter to activate the desired items.

So, in order to recall the digits of π in order, the person has to walk in the same path as when that person was storing the information (laying it down). It appears that this method led to the discovery of the so-called "Feynman Point"—a sequence of six consecutive 9's starting at the 762nd decimal place in π (Wells 1986: 51). Douglas Hofstadter (1985) was one of the first to notice the presence of these nines:

> I myself once learned 380 digits of π, when I was a crazy high-school kid. My never-attained ambition was to reach the spot, 762 digits out in the decimal expansion, where it goes "999999," so that I could recite it out loud, come to those six 9's, and then impishly say, "and so on!"

The point is called the Feynman Point after physicist Richard Feynman (Arndt and Hanel 2006: 3), who allegedly discussed it in a lecture. It is not certain, however, that Feynman ever made such an observation, since it is not mentioned either by Feynman himself or his biographers. The next sequence of six consecutive identical digits is again composed of 9's, starting at position 193,034. And the next distinct sequence of six consecutive identical digits—all 8's—starts at position 222,299, while strings of nine 9's next occur at position 590,331,982 and 640,787,382 (Wells 1986). These are suggestive patterns within π, but they do not occur recursively, and thus their discovery has not led to any definitive repeating pattern within π.

Another symptom of Pi-Mania can be seen in the quest to calculate π to larger and larger values. The first general computer, the Electronic Numerical Integrator and Computer (ENIAC) made it possible to calculate π to more than 2,000 digits in 1949. With a more sophisticated computer, in 2002 Yasumasa Kanada was able to compute π to 1.2 trillion decimal places, and in 2019, Emma Haruka Iwao, calculated π to a mind-boggling 31.4 trillion digits, or $π \times 10^{13}$ (Yee 2019). One can only guess on what surface we would be able to lay out the digits. Consider just the first 100 digits of π:

3.14159265358979323846264338327950288
4197169399375105820974944592307816
40628620899862803482534211706679 ...

Achieving such results was made possible, starting in the 1980s, with the development of new algorithms for computing π. One of these was devised by David and Gregory Chudnovsky in 1988, which they based on one of Ramanujan's π formulas. The Chudnovsky algorithm was used to calculate π to 2.7 trillion digits in 2009, 10 trillion in 2011, 22.4 trillion digits of π in 2016, 31.4 trillion digits in 2018, and a truly mindboggling 50 trillion digits in early 2020.

Why the quest to compute π to greater and greater lengths? In actual fact, most scientific applications only require π to several hundred digits. As Arndt and Hanel (2006) point out, 39 digits are sufficient to carry out most cosmological calculations, which is the accuracy level required to calculate the circumference of the observable universe to within one atom's diameter. Today's pursuit of π to longer and longer stretches of numbers is more about the ancient Pythagorean search for meaning in numbers and geometric forms than anything else. And, in fact, attempts to approximate π start in antiquity, growing exponentially after computers came onto the scene to help with these approximations in the 1950s, as the graph below shows.

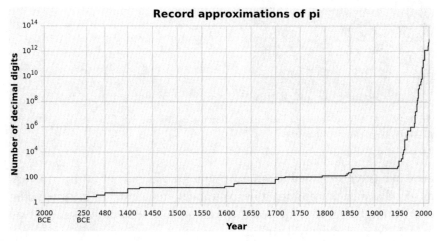

Pi approximations
SOURCE: WIKIMEDIA COMMONS

The obsession over π does not stop at the ingenious construction of piems or at feats of computation. It has led to the establishment of several calendar days dedicated to its celebration, where such constructions and feats are acknowledged and expanded. The most famous is Pi Day, on March 14, because in month-day format this is 3–14, which corresponds, of course, to the first digits of π (3.14). In 2009 the US House of Representatives put forth a non-binding resolution making March 14 officially recognized as Pi Day. The celebration starts at 1:59 PM, which when combined with the date makes 3.1459.

In addition to Pi Day, Pi Approximation Day is also celebrated by the most fervent aficionados of π on July 22, which is 22/7 in day/month format, given that 22/7 is a common approximation of π, traced back to Archimedes. Finally, Tau Day, which is also called Two Pi Day, is observed by some on June 28 (6/28) in honor of Tau or 2π, since 2π = 6.283185 ....

The first official Pi Day was organized in 1988 by physicist Larry Shaw at the San Francisco Exploratorium (his place of work). The celebration involved people marching around one of the circular spaces of the Exploratorium, and then eating fruit pies—a play on *pi* as a homophone of *pie* and on the fact that most pies are circular. Actual pies have been created on purpose for Pi Day, such as the one below created at Delft University of Technology on Pi Day, 2008:

Pie created at Delft University of Technology, March 14, 2008
SOURCE: WIKIMEDIA COMMONS

Participants are supposed to wear pi-themed attire, read books on π, watch movies that incorporate it into their plot, and become involved in various other activities that relate to π in some way. Logically, mnemonic feats are central to Pi Day, designed to break records.

A study by Raz, Packard, Alexander, Buhle, Zhu, and Peterson (2009) found that, reciting π actually enhances memory. The researchers looked at a subject who was adept at reciting the digits of π. Using functional magnetic resonance imaging (fMRI) and brain volumetric data, they found that while reciting the first 540 digits of π, the subject revealed increased activity in specific areas of the brain (the medial frontal gyrus and the dorsolateral prefrontal cortex). Pre-testing had indicated that the subject was of average intelligence and possessed normal procedural and implicit memory. Without going into details of the study, the implication is that memorizing π may have hidden neural benefits. Of course, this could relate to other types of mnemonic learning; nevertheless, the fact that π by itself may have benefitted the subject's memory is something that certainly requires future psychological consideration.

A study by Kaschube et al. (2010) appears to imply the π itself may be part of evolution. The researchers found that three distantly-related mammals shared a common organizing system for neurons in the visual cortex, which is characterized, remarkably, by a density closely approaching 3.14 (π). The appearance of π in this experiment is nothing less than amazing.

5   Epilogue

To distinguish between ancient forms of mythic stories and their resurgence in modern-day pop culture, semiotician Roland Barthes (1957) introduced the notion of *mythologies*. These are defined as contemporary versions of the ancient myths by allusion, inference, implication, or suggestion. This might plausibly explain why there are so many myths related to π in movies, programs, etc.

It is relevant to conclude this chapter with a final comment on Srinisava Ramanujan, characterized by Kanigel as *The Man Who Knew Infinity* (see also Murtry and Murtry 2013). Ramanujan devised his first theorems at age 13. By the age of 23 he was a recognized genius in the Indian mathematical community, even though he had no university education, having been rejected twice in the entrance exam for leaving unanswered all those questions that were not related to mathematics. As a consequence, he became strictly self-taught. He had one obsession that would follow him until the end of his life—the number π. He came up with hundreds of different ways of calculating approximate values of π, many of which are now used as part of number theory (Pickover 2005). In two of his notebooks, we find hundreds of pages of formulas and theorems about π. These have laid the theoretical foundations for devising computer algorithms that have calculated π to trillions of digits. Ramanujan's work is Pythagorean in the sense that it explores numbers in themselves as part of a hidden language of the universe. This is hardly an obsession; it is the dedication to exploring the many mysterious patterns in the world that mathematics makes viable.

As journalist Thomas Friedman (2007: 300) has aptly put it, the world today constitutes "a new age of numbers" in which "partnerships between mathematicians and computer scientists are bulling into whole new domains of business and imposing efficiencies in math." The contemporary arts and entertainment worlds have taken notice of mathematics, making it obvious that the theories of mathematicians are of great relevance to understanding life. In pop culture, everyone can earn a Doctor in Thinkology, no matter their mathematical training. As Mark Twain aptly wrote: "Intellectual work is misnamed; it is a pleasure, a dissipation, and is its own highest reward" (quoted in *A Connecticut Yankee in King Arthur's Court*, Chapter 28, 1889).

CHAPTER 5

Order and Chaos

> To study history means submitting to chaos and nevertheless re-
> taining faith in order and meaning.
>
> HERMANN HESSE (1877–1962)

∴

1 Prologue

The movie π: *Faith in Chaos* (1998), directed by Darren Aronofsky, raises fundamental philosophical questions about the relation between mathematics and
reality in true Pythagorean fashion—questions that Aronofsky sees as embedded in the ratio π. Why is something that is chaotic (a number with random
digits) describe the hidden Order of nature and the physical world? Is π a key
to unlocking the mystery of existence?

   A brilliant young mathematician, named Maximilian Cohen, teeters on the
brink of insanity as he searches for any elusive pattern in π's digits. His premise is that there must be a pattern in π, otherwise its appearance in natural
phenomena and human affairs would have no meaning. For the previous ten
years, Cohen was on the verge of unlocking the pattern concealed in the chaotic stock market indexes, with π appearing to be the key to unraveling it. So,
he concludes that the solution to figuring out the stock market is to uncover
the hidden pattern in π—the two problems, Cohen suspects, are one and the
same. As Cohen verges on a solution, an aggressive Wall Street firm, set on
financial domination of the stock market, and a Kabbalah sect, intent on unlocking the secret name of God hidden in their ancient holy text, the Torah,
which they believe may be concealed in the digits of π, approach him, as he
attempts to crack the code. In the end, Cohen does indeed discover the pattern
in π, but he throws away his solution irretrievably in a garbage can, and because of his unbalanced state of mind, cannot, or does not want to, reconstruct
his solution. This leaves viewers in a state of disquiet. A mystery produces a
kind of existential Angst that will linger until the mystery itself is unraveled—
and in this case it seems that it might never be solved.

© KONINKLIJKE BRILL NV, LEIDEN, 2021 | DOI:10.1163/9789004433397_006

Like the Pythagoreans, Aronofsky sees numbers as keys to our understanding of the universe. And, as Carl Sagan suggested metaphorically in *Contact* (previous chapter), there may indeed be a divine message within the digits of π for us to figure out. The two movies—*Contact* and *π: Faith in Chaos*—are fictional essays on our need to answer the mysteries of existence, encapsulated in the conundrum of π. Aronofsky's movie is, in effect, a filmic Pythagorean treatise on the relation between numbers and truth, suggesting overtly that mathematics is the language in which the book of nature is written. In his book *Il saggiatore* (1623), Galileo puts it as follows (cited in Popkin 1966: 65):

> Philosophy is written in this grand book—I mean the universe—which stands continually open to our gaze, but it cannot be understood unless one first learns to comprehend the language in which it is written. It is written in the language of mathematics, and its characters are triangles, circles, and other geometric figures, without which it is humanly impossible to understand a single word of it; without these, one is wandering about in a dark labyrinth.

As modern quantum physics attempts to develop a theory of the universe on the basis of increasingly abstract mathematics, and as AI produces algorithms that have an increasingly sophisticated ability to search for pattern in chaotic entities, it does not seem far-fetched to pursue Cohen's quest through advanced mathematics. It may be the only way to grasping and taming Chaos, called entropy by physicists. Entropy is the measure of the amount of disorder or randomness in a system, with randomness being the more probable state of the system. For instance, shuffling a deck of cards is likely to lead to a jumbled distribution of cards, not an ordered sequence. Randomness seems to be the default in all systems—human and natural. But the human mind cannot accept this state of affairs, and when confronted with randomness, such as the digits in π, it strives to search for Order within π nonetheless. What we vaguely call "meaning" is really Order.

This search is sometimes called an exercise in *organicism*, or the view that the universe and its parts constitute organic wholes. This view goes back to the origins of mathematics as the key to describing the world as a whole with interconnected parts—which has been named, as discussed, *musica universalis* theory. The basic forms of reality are thus immanent in the universe, and these are mirrored in mathematical laws. The problem is that π emerges from an "organic (holistic) structure," the circle, yet π itself seems to defy the laws of organicism, that is, laws of order—hence an existential paradox.

In this final chapter, the method of hermeneutic geometry is applied to philosophical questions such as those raised by Aronofsky's movie, but which go right back to the origins of mathematics. The movie is, overall, a fictional essay in hermeneutic geometry.

2   Cohen's Dilemma

The dilemma that Cohen faces is a Pythagorean one that entails a dichotomy between mysticism and science, as emphasized in one of the dialogues between Cohen and his university mentor, Sol Robeson, in which Sol warns Cohen that: "As soon as you discard scientific rigor, you are no longer a mathematician, you are a numerologist" (All citations from: https://en.wikiquote.org/wiki/Pi). But Cohen ignores the critique, in the same way that he ignored a warning from his mother in childhood: "When I was a little kid, my mother told me not to stare into the sun. So once when I was six, I did." The subtext is that humans have a thirst to unravel mysteries even if this might involve personal risk: "It's fair to say that I'm stepping out on a limb, but I am on the edge and that's where it happens."

Throughout the movie Cohen articulates a set of three principles that motivate his desperate search for the pattern concealed in π—principles that recycle Pythagorean ones in contemporary terms:

One: Mathematics is the language of nature.
Two: Everything around us can be represented and understood through numbers.
Three: If you graph the numbers of any system, patterns emerge.
Therefore: There are patterns everywhere in nature.

But understanding these patterns is another matter—very much like the universe itself, the more advanced we become and as our picture of π grows larger, the more its mysteries grow. Indeed, we may have to face the fact that solving the mystery of π may be beyond human capacity. As Pickover (2005: 85) aptly observes, alluding to Sagan's *Contact*:

> How can the sum of an infinite series be connected to the ratio of the circumference of a circle to its diameter? This demonstrates one of the most startling characteristics of mathematics—the interconnectedness of, seemingly, unrelated ideas. It is as if there existed some great landscape of meta-mathematics, and we are only seeing the peaks of mathematical

mountains above the valley fog. Could some intelligent aliens have such superior brains that the fog is lifted from their eyes and all the interconnections become apparent?

A major theme put forth by Aronofsky's movie is the question of the link between mathematics and the nature of consciousness. In a relevant dialogue between Max and Sol, this question is broached as follows, after Max reports that his computer (called rather appropriately Euclid) had crashed as it was trying to solve the code in π:

[SOL]   My guess is that certain problems cause computers to get stuck in a particular loop. The loop leads to meltdown, but just before they crash they ... they become "aware" of their own structure. The computer has a sense of its own silicon nature and it prints out its ingredients.

[MAX]   The computer becomes conscious?

[SOL]   In some ways ... I guess.

[MAX]   Studying the pattern made Euclid conscious of itself. Before it died it spit out the number. That consciousness is the number.

[SOL]   No, Max, it's only a nasty bug.

[MAX]   It's more than that!

[SOL]   No it's not! It's a dead end, there is nothing there.

[MAX]   It's a door Sol, a door.

[SOL]   A door in front of a cliff. You're driving yourself over the edge. You need to stop.

[MAX]   You were afraid of it. That's why you quit.

[SOL]   Max, I got burnt.

[MAX]   C'mon, Sol.

[SOL]   It caused my stroke.

[MAX]   That's bullshit. It's mathematics, numbers, ideas. Mathematicians are supposed to be out on the edge. You taught me that!

[SOL]   Max, there's more than math! It's death Max!

Cohen cannot let go of his quest to unravel the meaning in π, because the alternative might be that life is without meaning. Why is there something rather than nothing? As implied in the repartee above, the "something" emerges because of consciousness; without it the universe would not exist in an ontological sense. But even if Cohen found the meaning of π, it might still leave him asking why there is something rather than nothing. Darwin (1859) had argued that consciousness was an organic principle of life, rather than as some "animal spirit," as Descartes (1633) called it, and thus a survival mechanism. But, a subtext in Aronofsky's movie is that evolution alone cannot explain consciousness. It tells us nothing about why humans create their meaningful experiences and pose the questions they do about life, even at the risk of losing their minds—as does Cohen by the end of the movie.

3   Chaos Theory

The founding of mathematics by Pythagoras was to pursue what Cohen does in the movie—to seek the Order in what otherwise would be the Chaos. Cohen's dilemma is a human one—life and existence cannot be random; it must have Order for it to have any meaning whatsoever. The problem arises when mathematics itself produces randomness, such as π.

The systematic study of randomness in mathematics came to the forefront with probability theory, which showed that quantities cannot be studied in absolutist terms, but relative to the situation in which they exist. There are now even algorithms, called Random Number Generators (RNG s), that are devised to generate sequences of numbers or symbols that lack any pattern, or at least, appear to be random. RNG s have had practical applications, as for instance in lotteries and PIN numbers. Investigating and even taming randomness has been an overriding goal of mathematics since the beginning. It has even produced a branch that studies Chaos itself—called Chaos Theory. This started at about the same time as fractal geometry. As we saw (Chapter 2), fractal geometry deals with complex shapes, which consist of small-scale and large-scale structures that resemble one another. Fractals describe many natural shapes, such as coastlines or branching trees, effectively. Although fractal structures seem irregular, they are based on an archetypal organizing principle—iteration. Chaos theory aims to unravel iterative structure in seemingly random phenomena, such as the stock market—which is what Cohen was attempting to do.

While most science deals with regular phenomena like gravity, electricity, or chemical reactions, Chaos Theory deals with phenomena that are impossible to predict or control entirely, like turbulence, our brain states, and so on. These are described, as discussed, by fractal mathematics. Having grasped the chaotic, fractal nature of our world is a remarkable achievement, with many practical consequences. For example, by understanding the fractal dynamics of the atmosphere, a pilot can steer a balloon to a desired location. By understanding that our social and economic systems are interconnected in some fractal way, we might be able to avoid actions that are detrimental to our long-term well-being.

Chaos Theory was founded officially by French mathematician Henri Poincaré (see the various essays in Skiadas 2018); its premise is that there are patterns hidden in random events. Already in the 1960s, simplified computer models demonstrated that there was a hidden structure in the seemingly chaotic patterns of the weather. When these were plotted in three dimensions they revealed a butterfly-shaped fractal set of points. So, the question arises if a Chaos Theory approach to π might reveal something that has not been noticed before? This is a rhetorical question, needless to say, but arguably not a trivial one. Through formulas and their graphs chaotic phenomena often reveal their structure. It is an implicit premise in π: *Faith in Chaos*, as can be seen by Cohen's set of principles above, which he repeats throughout the movie—namely, that mathematics is the language of nature and that everything is number, so that if you graph the numbers patterns will emerge.

Chaos Theory explores feedback loops, repetition, self-similarity, iterative fractals, and self-organization. A chaotic system may have either sequences that exactly repeat themselves, producing periodic behavior, or random sequences. Much of the mathematics of Chaos Theory involves the iteration of formulas, which would be impractical to do by hand. It may be the key to deciphering π, as Aronofsky's movie alluded in the title itself—*Faith in Chaos*. The problem is, however, that π has not given up its secrets yet, and it may never do so (if there are secrets in it). Randomness is the opposite of recursion. But the two may well be in a dynamic interrelationship of which we still know very little. An example may be the M-Set (Chapter 2), which results from a fractal recursion (iterative) formula, as we saw, whose graph is the M-Set. Today, computers are used to produce graphs of this set to magnify any section of the fractal by making more and more calculations on that section. The magnifications have revealed an infinitude of repeating patterns. Could this technique reveal something about π as well?

The paradox is that π is an incommensurable irrational number, yet provides a basis for measuring circumferences, volumes, and areas of spheres. Moreover, it can be located on the number line:

Pi on the number line

As mentioned (Chapter 1), the Pythagoreans discovered the first irrational number, $\sqrt{2}$, which they kept secret (Fritz 1945). But the secret soon came out, and Greek mathematicians initially named this incommensurable ratio *alógos* (without *lógos*). The discovery of incommensurable ratios introduced the distinction into mathematics between discrete and continuous. Zeno questioned the notion that quantities are discrete and composed of a finite number of units with his brilliant paradoxes. Previously, mathematicians believed that "whole numbers represent discrete objects, and a commensurable ratio represents a relation between two collections of discrete objects" (Kline 1972: 32). Zeno showed, on the other hand, that quantities "in general are not discrete collections of units; this is why ratios of incommensurable numbers appear ... they are, in other words, continuous" (Kline 1972: 34).

Eudoxus of Cnidus was among the first to formalize a theory of ratios that took into account commensurable as well as incommensurable quantities (Kline 1972: 48). He was the one who first made the key distinction between magnitude and number—the former "was not a number but stood for entities such as line segments, angles, areas, volumes, and time which could vary, as we would say, continuously," while the latter "jumped from one value to another, as from 4 to 5" (Kline 1972: 48). Numbers are indivisible units, magnitudes are, instead, infinitely reducible. Eudoxus was thus able to account for both commensurable and incommensurable ratios by defining a ratio in terms of its magnitude, and a proportion as an equality between two ratios, laying the "logical foundation for incommensurable ratios" (Kline 1972: 49). The incommensurability issue was taken up by Euclid in Book X, Proposition 9 of his *Elements*.

It has been found that π is a transcendental number, defined as any non-algebraic number—any number that can be a root in an equation. In the equation below x is algebraic, because it has two values (roots of the equation):

$$x^2 + x = 6 \ (x = 2, -3)$$

Some irrational numbers (such as $\sqrt{2}$) are algebraic—recall that $\sqrt{2}$ is the solution to the equation $x^2 = 1^2 + 1^2$ (Chapter 1). There are many numbers that are not algebraic, called transcendental. One of the first to discuss these was Joseph Liouville (1844), who came up with the following transcendental number, now called the Liouville constant:

0.1100010000000000000000000100 ... ...

This number has the following features:
- The digit is 1 if it is $n!$ places after the decimal, and 0 otherwise. For example for *1!*, the 1 is in the first position; for *2!*, 1 is in the second position; for *3!* (= 6), the 1 is the sixth position; for *4!* (= 24), the 1 is in the 24th position;

and so on. A factorial, represented with an exclamation mark, is the product of an integer and all the integers below it; so, factorial four (4!) is equal to: $4 \times 3 \times 2 \times 1 = 24$.

- It is irrational, but not the root of any polynomial equation and so is not algebraic.
- Liouville had successfully constructed the first provable transcendental number.

It was in 1873 when the first "non-constructed" number proved to be transcendental; that number was e, defined as the limit of the expression $(1 + 1/n)^n$ as n becomes large without bound. Its limiting value is approximately 2.7182818285. In 1882, Ferdinand von Lindemann proved that π was transcendental. Since then, it has become clear that most numbers are transcendental. But this type of discovery takes us right back to Cohen's dilemma. Why is π, a transcendental number, so ubiquitous and, like many other transcendentals, incommensurable? Why is their Chaos in Order? As Yann Martel remarks in his novel *Life of Pi* (Chapter 4), all things "contain a measure of madness that moves them in strange, sometimes inexplicable ways."

4   Order and Chaos

Chaos is something that we find averse. The digits in π are seemingly chaotic, yet, as Aronofsky's movie suggests, there simply must be a pattern hidden within them, or else our sense of Order may break down. But unlike the taming of many aspects of randomness by Chaos Theory, the randomness in the digits cannot be tamed in any foreseeable way.

In his *Theogony*, the Greek poet Hesiod (eighth century BCE) states that Chaos generated Earth, called Gaea, from which arose the starry, cloud-filled Heaven (Hesiod 1987). In another Greek myth, Chaos was portrayed as the formless matter from which the Cosmos, or harmonious Order, was created. In both myths, it is obvious that the ancients felt that Order emerged from the Chaos. This feeling continues to reverberate in contemporary humans. For some truly mysterious reason, we require, like the ancients, that there be Order within apparent Disorder. The search has been a central motivating force of human history. And it will continue to be so as long as humans are around. But it may also affect us negatively. As Polster and Ross (2012) point out about Aronofsky's movie, when our "faith" in Order breaks down, we are left in an emotional quandary that may have serious consequences for our sanity, suggesting that Order and Chaos may be in a tug of war within the brain, rather than in the world—a theme that saturates Aronofsky's movie.

The solution to π seems to be hidden in a 216 digit decimal number uncovered by Cohen that the Kabbalah mystics believe hides the name of God. But this number wreaks havoc with whoever gets close to it. For example, Cohen's computer crashes while analyzing patterns in the stock market and the Torah that conceal this number; he suffers debilitating migraines as he contemplates the solution; and his mentor Sol Robeson suffered two strokes before him, dying from the second one. Cohen ends up lobotomizing himself by drilling into his brain to eradicate all traces of the number. The moral lesson of the movie is an obvious one—π emerges from Chaos and should be left there, otherwise it will wreak havoc in everyone who tries to decipher its "syntax," as Cohen says in an episode, claiming that the meaning of the 216 digits is not to be found in them as such but in how they are combined together. By analogy, consciousness itself stems from the combination of separate thoughts, much like the meaning of a sentence inheres not in its separate words, but in the syntax that unites them.

The connection of the 216 digits to mysticism on the part of the religious sect in the movie alludes to the art of the ancient Israelites called *gematria*, or the view that the letters of any word could be translated into digits and rearranged to form a number that contained a secret (sacred) message. The earliest recorded use of gematria was by King Sargon II of Babylon in the eighth century BCE, who built the wall of the city of Khorsabad exactly 16,283 cubits long because this was the numerological value of the letters in his name. The same type of belief can be found in the origin of anagrams, which were thought in antiquity to be the linguistic vehicles used by the divinities to communicate with mortals. Many ancient prophets were essentially "anagrammatists" who interpreted this purported heavenly form of language. Soothsayer status was often attained by those who claimed to possess knowledge of anagrams. In the third century BCE, the Greek poet and prophet Lycophron made a profession of devising anagrams of the names of the members of the Hellenistic King Ptolemy II's court in Egypt, as a basis for divining each person's character and destiny. For this, he became widely known and sought out as a soothsayer.

Throughout the ancient world, it was commonly believed that personal names were anagrams foretelling the fates of individuals. People would often wear amulets with anagrams of their names on them to ward off evil. All this may be connected with the belief that writing itself has a divine origin. The etymology of the term *hieroglyph*, to describe early Egyptian writing, bears witness to this belief—*hieroglyphic* derives from Greek *hieros* "holy" and *glyphein* "to carve." In their origins writing systems across the world were deemed to have divine power, and the myths of many cultures confirm this by attributing the origin of writing to deities—the Cretans to Zeus, the Sumerians to Nabu, the Egyptians to Toth, the Greeks to Hermes, and so on.

In both scenarios—the stock market and the Kabbalah sect—there is a search for the meaning of numbers beyond numeration; in this scenario π is the numerical equivalent of an anagram: Does it reveal mystical knowledge if its syntax is reordered and reconstituted?. This is articulated by Sol in the movie as follows, making a relevant commentary on Cohen's state of mind, as they play the Japanese board game of Go:

[SOL]   The ancient Japanese considered the Go board to be a microcosm of the universe. Although, when it is empty it appears to be simple and ordered, the possibilities of game play are endless. They say that no two Go games have ever been alike. Just like snowflakes. So, the Go board actually represents an extremely complex and chaotic universe and that's the truth of our world, Max. It can't be easily summed up with math. There is no simple pattern.

[MAX]   But as a Go game progresses, the possibilities become smaller and smaller. The board does take on order. Soon, all moves are predictable.

[SOL]   So, so?

[MAX]   So, maybe, even though we're not sophisticated enough to be aware of it, there is a pattern, an order, underlying every Go game. Maybe that pattern is like the pattern in the stock market, the Torah. This two sixteen number.

[SOL]   This is insanity, Max.

[MAX]   Or maybe it's genius. I have to get that number.

[SOL]   Hold on, you have to slow down. You're losing it, you have to take a breath. Listen to yourself. You're connecting a computer bug I had, with a computer bug you might have had, and some religious hogwash. If you want to find the number two sixteen in the world, you'll be able to find it everywhere. Two hundred sixteen steps from your street corner to your front door. Two hundred sixteen seconds you spend riding on the elevator. When your mind becomes obsessed with anything, you will filter everything else out and find that thing everywhere. Three hundred and twenty, four hundred and fifty, twenty-three. Whatever! You've chosen two hundred sixteen and you'll find it everywhere in nature. But Max, as soon as you discard scientific rigor, you are no longer a mathematician. You are a numerologist.

In the starting sequence of the movie, we see the decimal expansion of π scrolling on the screen, foreshadowing what is going on in Cohen's brain during one of his migraine episodes. It also suggests that these digits are literally everywhere, but their meaning is in our minds and thus we can find them everywhere, as Sol emphasizes.

Yet there may be some archetypal feature at work that is not yet evident in π. Early on, the archetypal concept of cycles comes up as a guiding thought in Cohen's search, applying it to the stock market:

EVIDENCE: the cycling of disease epidemics, the wax and wane of Caribou populations, sunspot cycles, the rise and fall of the Nile. So what about the stock market? The universe of numbers that represents the global economy. Millions of human hands at work. Billions of minds, a vast network screaming with life, an organism, a natural organism.

MY HYPOTHESIS: Within the stock market there is a pattern as well, right in front of me, hiding behind the numbers. Always has been.

Cohen pays homage to Pythagoras in another episode, thus linking the theme of the movie to Pythagoreanism explicitly:

Remember Pythagoras. Mathematician, cult leader, Athens, ca. 500 B.C.E. Major belief: The universe is made of numbers. Major contribution: the Golden Ratio. Best represented geometrically as the Golden Rectangle ... Pythagoras loved this shape for it is found everywhere in nature: the nautilus shell, rams horns, whirl pools, tornados, our fingerprints, our DNA, and even our Milky way.

In the final scene of the movie, we see Cohen sitting on a bench next to a little girl who lives in the same apartment complex in which he resides and who always asks him to help her solve simple problems in arithmetic. We see Cohen giving in to the little girl, engaging her in dialogue over her arithmetical problems. We are left with the sense that Cohen has come to the conclusion that pursuing the unsolvable may be a symptom of existential insanity, as his mentor always empathized to him. This seems to restore his peace of mind, but it leaves us, the viewers, in a state of permanent suspense that any unsolved mystery invariably produces.

5   Epilogue

Discovered at the dawn of civilization, π is among the oldest manifestations of human ingenuity. It has been both a guide to understanding facets of the universe, and also a conundrum, since it does not, itself, seem to possess any hidden meaning. It is thus a metaphor for who we are—a species that seeks Order in the Chaos and will do so relentlessly, as Aronofsky's movie suggests. That is how the human imagination operates—it is a restless seeker of meaning.

We rare still seeking for patterns in π, using algorithms to help us out, as discussed in the previous chapter. A neural network model may provide relevant insights, since it can mimic the many separate algorithms for producing the digits of π, searching for pattern among the many formulas devised for computing it. But, like Cohen's unyielding search, it is unlikely that AI will provide any deep revelation about π. Computer algorithms can solve complex problems, but start having difficulty as the degrees of complexity increase. The issue of complexity raises the related issue of decidability. This is the notion that there would be no point in tackling a complex problem that may turn out not to have a solution or that might be so complex that it would take an enormous amount of computer time to solve, thrusting it into a loop (as does π). Perhaps the search for pattern in π is influenced by the fact that we lay out our digits in a linear style, which leads us to interpret it as have sequential structure. Maybe, like the pie created for Pi Day (Chapter 4), laying them out in a different way, perhaps in some way that resembles the structure of circularity or curvature, some pattern might crystallize. This is only speculation, of course; the point is that the way we represent and symbolize facts influences how we perceive them and thus what kind of information we may be able to extract from them.

In reality, there are no linear paths that lead to definitive answers. In Carroll's *Alice's Adventures in Wonderland* (Carroll 1865 71–72), Alice asks the Cheshire cat: "Would you tell me please, which way I ought to go from here?" The cat's answer is simple, yet revealingly insightful: "That depends a good deal on where you want to get to," to which Alice responds with "I don't much care where." The shrewd cat's rejoinder to Alice's response applies to the search for pattern in π: "It doesn't matter which way we go."

All that can be said is that π may well be an elusive bit of evidence of a theory of the world that is lurking around somewhere, but that seems to evade formulation. From the Pythagorean practice of giving sacrifice to the gods for mathematical discoveries to the seventeenth century practice on the part of the Japanese of giving *sangaku* (the Japanese word for "mathematical tablet") to the spirits for discovering mathematical proofs, there seems to be a universal feeling in us that we can never understand existence in its totality, only in bits and pieces.

In this book, the methods of hermeneutic geometry have been suggested to explore ideas such as those found in Aronofsky's movie and which reach right back to Pythagoras. But even the method of hermeneutics really does not penetrate the substance of the enigma at hand. Nor does it really answer the question of the interconnection between mathematical models and their serendipitous appearance in the world. So we are left with the same kinds of questions with which we started off this book and which Cohen constantly articulates throughout the movie: Why does mathematics work to explain the physical world? But why, then, do conundrums within mathematics, such as π and other irrational numbers exist as seemingly random entities that nonetheless describe aspects of our universe? As Clawson (1999: 284) has suggested, mathematics might even explain the laws of unknown universes: "Certain mathematical truths are the same beyond this particular universe and work for all potential universes."

Kurt Gödel (1931) showed rather matter-of-factly that there never can be a consistent system of propositions that can capture all the truths of mathematics, making it clear, in effect, that the makers of the statements could never extricate themselves from making them. Gödel's proof demonstrated convincingly that the exploration of mathematical truth would go on forever as long as humans were around. The complete map of the mathematical realm will never be drawn. Like other products of the imagination, mathematics lies within the minds of humans and thus can only give us partial glimpses of reality.

6   Final Remarks

Four themes have guided the present foray into the meaning of π. The first one is that of mystery, captured eloquently by physicist Richard Feynman as follows (in Feynman and Robbins 1999 177):

> As a youth, fiddling in my home laboratory, I discovered a formula for the frequency of a resonant circuit which was $2\pi \times \mathrm{sqrt}(LC)$ where L is the inductance and C the capacitance of the circuit. And there was π, and where was the circle? ... I still don't quite know where that circle is, where that π comes from.

Rather than to see π as connected just to the circle, it is more appropriate to link it to archetypes such as circularity, curvature, infinity, and others where discrete commensurability does not apply. These are everywhere in nature and life. Pi is only a key to unlocking their structure, but it tells us nothing about why archetypes are around in the first place.

A second theme has been that π has become an obsession for some—as emblemized by Aronofsky's film—and incorporated in human practices and domains such as art, architecture, movies, TV programs, etc. It has been the inspiration of a multimedia installation in the vicinity of the Karlsplatz in Vienna, located in the Opernpassage between the entrance to the subway and the stop in Secession near the Naschmarkt. The installation was designed by Canadian artist Ken Lum from Vancouver which he completed in December of 2006. It consists of statistical information and a representation of π to 478 decimal places:

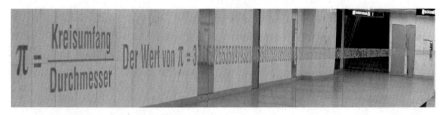

Pi in Vienna near Karlsplatz, Secession
SOURCE: WIKIMEDIA COMMONS

The third theme has been that π is a paradox—a number resulting for a perfect geometric form, the circle, yet possessing randomness (non-perfection) in it digits. As Arthur Koestler insightfully remarked in his book *The Sleepwalkers* (1959):

> These geometrical ratios are the pure harmonies which guided God in the work of Creation; the sensory harmony which we perceive by listening to musical consonances is merely an echo of it. But that inborn instinct in man which makes his soul resonate to music, provides him with a clue to the nature of the mathematical harmonies which are at its source. The Pythagoreans had discovered that the octave originates in the ratio 1:2 between the length of the two vibrating strings, the fifth in the ratio of 2:3, the fourth in 3:4, and so on. But they went wrong, says Kepler, when they sought for an explanation of this marvellous fact in occult number-lore. The explanation why the ratio 3:5, for instance, gives a concord, but 3:7 a discord, must be sought not in arithmetical, but in geometrical considerations.

Pi describes the circle yet is incommensurable and thus does not seem to fit in with the Pythagorean view of Order. Yet, it shows up in calculations of the orbits, as Kepler showed, thus indirectly (and paradoxically) substantiating the Pythagorean *musica universalis* theory. The paradox becomes even more compelling, as Aronofsky's movie emphasizes, by the fact that π crops up not only in mathematics, but also in natural and physical structures and in human affairs. Could it be the key that unlocks the meaning of the universe? The movie appears to answer this question in the affirmative, but the 216-digit number that Cohen discovers is lost to us forever—implying that the mystery of existence may be beyond humans to unravel. As the nineteenth century mathematician Augustus De Morgan so aptly put it, π may be a key because it "comes in at every door and window, and down every chimney" (quoted in Sophia Elizabeth De Morgan, *Memoir of Augustus De Morgan* 1882: 46).

The fourth theme has been that geometry is a hermeneutic science, since it allows us to investigate the linkage that exists between geometrical constructs and reality. The aim of hermeneutic geometry is encapsulated indirectly by mathematician and engineer Antranig Basman as part of a series of quotes collected for Pi Day: "Pi is not just a collection of random digits; pi is a journey; an experience; unless you try to see the natural poetry that exists in pi, you will find it very difficult to learn."

It is appropriate to conclude this foray into the meaning of π by citing the words of one of the creators of the π algorithm discussed in the previous chapter, mathematician David Chudnovsky, since it summarizes the theme of this book eloquently: "Exploring pi is like exploring the universe" (cited in *Times of Malta*).

References

Abbott, Edwin A. (1884) [2002]. *Flatland: A Romance of Many Dimensions. Flatland: A Romance of Many Dimensions.* Introduction and Notes by Ian Stewart. New York: Basic Books.

Adam, John A. (2004). *Mathematics in Nature: Modeling Patterns in the Natural World.* Princeton: Princeton University Press.

Agarwal, Ravi P., Agarwal, Hans, and Sen, Syamal K. (2013). Birth, Growth and Computation of Pi to Ten Trillion Digits. *Advances in Difference Equations.* http://www.advancesindifferenceequations.com/content/2013/1/100.

Al-Khalili, Jim (2012). *Paradox: The Nine Greatest Enigmas in Physics.* New York: Broadway.

Alexander, James (2012). On the Cognitive and Semiotic Structure of Mathematics. In: Mariana Bockarova, Marcel Danesi, and Rafael Núñez (eds.), *Semiotic and Cognitive Science Essays on the Nature of Mathematics*, 1–34. Munich: Lincom Europa.

Aristotle (350 BCE). *On the Heavens.* classics.mit.edu/Aristotle/heavens.2.

Aristotle (1952/350 BCE). *Poetics*, in *The Works of Aristotle*, Vol. 11, ed. by W. D. Ross. Oxford: Clarendon Press.

Aristotle (1999/350 BCE). *Metaphysics*, trans. by J. Sachs. Santa Fe: Green Lion Press.

Aristotle (2015/350 BCE). *Physics*, trans. by R. P. Hardie and R. K. Gaye. Aeterna Press. http://classics.mit.edu/Aristotle/physics.4.iv.html.

Aristotle (2016/350 BCE). *On Interpretation.* CreateSpace Independent Publishing Platform.

Arndt, Jörg and Haenel, Christoph (2006). *Pi Unleashed.* New York: Springer-Verlag.

Asimov, Isaac (1959). *Realm of Numbers.* Boston: Houghton Mifflin.

Asimov, Isaac (1960). *Realm of Measure.* Boston: Houghton Mifflin.

Asimov, Isaac (1961). *Realm of Algebra.* Boston: Houghton Mifflin.

Asimov, Isaac (1964). *Quick and Easy Math.* Boston: Houghton Mifflin.

Asimov, Isaac (1965). *An Easy Introduction to the Slide Rule.* Boston: Houghton Mifflin.

Asimov, Isaac (1973). *How Did We Find Out About Numbers?* London: Walker.

Askew, Mike and Ebbutt, Sheila (2011). *From Pythagoras to the Space Race: The ABC of Geometry.* London: New Burlington Books.

Auburn, David (2000). *Proof.* Dramatists Play Service, Inc.

Augustine, Saint (1887). *De doctrina christiana. Patrologia latina, migne tomus XXXIV.* Paris: Hachette.

Baker, Alan (1990). *Transcendental Number Theory.* Cambridge: Cambridge University Press.

Ball, W. W. Rouse (1905) [1972]. *Mathematical Recreations and Essays*, 12th edition, revised by H. S. M. Coxeter. Toronto: University of Toronto Press.

Banks, Robert S. (1999). *Slicing Pizzas, Racing Turtles, and Further Adventures in Applied Mathematics*. Princeton: Princeton University Press.

Barnsley, Michael (1988). *Fractals Everywhere*. Boston: Academic.

Barr, Stephen (1964). *Experiments in Topology*. New York: Dover.

Barthes, Roland (1957). *Mythologies*. Paris: Seuil.

Barthes, Roland (1981) Theory of the Text. In: R. Young (ed.), *Untying the Text*, 31–47. London: Routledge.

Baum, L. Frank (1900). *The Wonderful Wizard of Oz*. Chicago: G. M. Hill.

Bayley, Melanie (2009). Alice's Adventures in Algebra: Wonderland Solved. *New Scientist*, Issue 2379.

Bear, Greg (1985). *Eon*. New York: Open Road.

Beckmann, Petr (1971). *A History of π*. New York: St. Martin's.

Beekes, Robert (2009). *Etymological Dictionary of Greek*. Leiden: Brill.

Bell, Eric Temple (1940). *The Development of Mathematics*. New York: Dover.

Bellos, Alex (2015). He Ate All the Pi: Japanese Man Memorises π to 111,700 Digits. *The Guardian*. theguardian.com/science/alexs-adventures-in-numberland/2015/mar/13/pi-day-2015-memory-memorisation-world-record-japanese-akira-haraguchi.

Berger, John. (1965). *The Success and Failure of Picasso*. Harmondsworth: Penguin Books.

Bergin, Thomas G. and Fisch, Max (1984). *The New Science of Giambattista Vico*. Ithaca: Cornell University Press.

Berlinski, David (2000). *The Advent of the Algorithm*. New York: Harcourt.

Berlinski, David (2013). *The King of Infinite Space: Euclid and His Elements*. New York: Basic Books.

Black, Max (1962). *Models and Metaphors*. Ithaca: Cornell University Press.

Blatner, David (1997). *The Joy of Pi*. Harmondsworth: Penguin.

Boole, George (1854). *An Investigation of the Laws of Thought*. New York: Dover.

Borel, Émil (1909). Le continu mathématique et le continu physique. *Rivista di Scienza* 6: 21–35.

Borgmann, Dmitri (1965). *Language on Vacation*. New York: Scribner's.

Bovill, Carl (1996). *Fractal Geometry in Architecture and Design*. Boston: Birkhauser.

Brahe, Tycho (1602). *Astronomiae Instauratae Progymnasmata*. Frankfurt: Godefridum Tampachium.

Brann, Eva (2011). *The Logos of Heraclitus*. Philadelphia: Paul Dry Books.

Brauen, Martin (1997). *The Mandala, Sacred Circle in Tibetan Buddhism*. London: Serindia Press.

Bremmer, Jan (2014). *Initiation into the Mysteries of the Ancient World*. Berlin: Walter de Gruyter.

Bronowski, Jacob (1973). *The Ascent of Man*. Boston: Little, Brown, and Co.

Brown, Tony (2001). *Mathematics Education and Language: Interpreting Hermeneutics and Post-Structuralism*. Dordrecht: Kluwer.

Buffon, Georges-Louis Leclerc (1777). Essai d'arithmétique morale. *Histoire Naturelle, Générale et Particulière, Supplément* 4: 46–123.

Bunt, Lucas, Jones, Phillip, and Bedient, Jack (1976). *The Historical Roots of Elementary Mathematics*. New York: Dover.

Burger, Dionys. (1995) [1965]. *Sphereland: A Fantasy about Curved Spaces and an Expanded Universe*, trans. Coreneli J. Rheinbolt. New York: Harper Collins.

Burkert, Walter (1972). *Lore and Science in Ancient Pythagoreanism*. Cambridge, MA: Harvard University Press.

Byrne, Rhonda (2006). *The Secret.* Atria Publishing Group.

Cajori, Florian (1916). *William Oughtred: A Great Seventeenth=Century Teacher of Mathematics.* Chicago: The Open Court Publishing Co.

Campbell, Joseph and Moyers, Bill (1988). *The Power of Myth.* New York: Doubleday.

Carpenter, Rhys (1921). *The Esthetic Basis Of Greek Art: Of The Fifth And Fourth Centuries B.C.* Bryn Mawr: Bryn Mawr Press.

Carroll, Lewis (1865). *Alice's Adventures in Wonderland.* London: Macmillan.

Carroll, Lewis (1871). *Through the Looking-Glass, and What Alice Found There*. London: Macmillan.

Ceulen, Ludolph van (1596). *Vanden Circkel.* Tot Delft.

Chace, Arnold Buffum (1979). *The Rhind Mathematical Papyrus: Free Translation and Commentary with Selected Photographs, Transcriptions, Transliterations and Literal Translations*. Reston: National Council of Teachers of Mathematics.

Changeux, Pierre (2013). *The Good, the True, and the Beautiful: A Neuronal Approach.* New Haven: Yale University Press.

Chartier, Tim (2014). *Math Bytes*. Princeton: Princeton University Press.

Chudnovsky, David and Chudnovsky, Gregory (1988). Approximation and Complex Multiplication according to Ramanujan. In: *Ramanujan Revisited: Proceedings of the Centenary Conference*, 375–472. Boston: Academic Press.

Clark, Kenneth (1956). *The Nude: A Study in Ideal Form*. Princeton: Princeton University Press.

Clark, Michael (2007). *Paradoxes from A to Z*. London: Routledge.

Clarke, Arthur C. and Baxter, Stephen (2003). *Time's Eye*. New York: Ballantine Books.

Clawson, Calvin C. (1999). *Mathematical Mysteries: The Beauty and Magic of Numbers*. Cambridge, MA: Perseus Books.

Cooke, Roger L. (2011). *The History of Mathematics: A Brief Course*. New York: John Wiley & Sons.

Courant, Richard and Robbins, Herbert (1941). *What Is Mathematics? An Elementary Approach to Ideas and Methods*. Oxford: Oxford University Press.

Cox, Brian and Forshaw, Jeff (2011). *The Quantum Universe*. Boston: DaCapo Press.

Coxeter, H. S. M. (1942). *Non-Euclidean Geometry*. Toronto: University of Toronto Press.

Coxeter, H. S. M. (1948). *Regular Polytopes*. London: Methuen.

Crilly, Tony (2011). *Mathematics*. London: Quercus.

Dancy, R. M. (2007). *Plato's Introduction of Forms*. Cambridge: Cambridge University Press.

Danesi, Marcel (2007). *The Quest for Meaning: A Guide to Semiotic Theory and Practice*. Toronto: University of Toronto Press.

Danesi, Marcel (2009). *X-Rated: The Role of Mythic Symbolism in Popular Culture*. New York: Palgrave.

Danesi, Marcel (2013). *Discovery in Mathematics*. Munich: Lincom Europa.

Danesi, Marcel (2018). *Ahmes' Legacy: Puzzles and the Mathematical Mind*. New York: Springer.

Danesi, Marcel (ed.) (2019). *Interdisciplinary Perspectives on Mathematical Cognition*. Boston: Springer.

Dantzig, Tobias (2005). *Number: The Language of Science*. New York: Plume.

Darwin, Charles (1859). *The Origin of Species*. New York: Collier.

Davies, Jamie A. (2013). *Mechanisms of Morphogenesis*. New York: Academic.

Davis, Paul J. and Hersh, Reuben (1986). *Descartes' Dream: The World according to Mathematics*. Boston: Houghton Mifflin.

Dehaene, Stanislas (1997). *The Number Sense: How the Mind Creates Mathematics*. Oxford: Oxford University Press.

De Morgan, Sophia Elizabeth (1882). *Memoir of Augustus De Morgan*. London: Longmans, Green, and Company.

Descartes, René (1633) *De Homine*. Amsterdam: Elsevier.

Devlin, Keith (2005). *The Math Instinct*. New York: Thunder's Mouth Press.

Devlin, Keith (2010). The Hidden Math Behind *Alice in Wonderland*. Mathematical Association of America. https://www.maa.org/external_archive/devlin/devlin_03_10.html.

Devlin, Keith and Lorden, Gary (2007). *The Numbers behind Numb3rs: Solving Crime with Mathematics*. New York: Plume.

Dewdney, A. K. (2001) [1984]. *The Planiverse: Computer Contact with a Two-Dimensional World*. New York: Copernicus Books.

Dilthey, Wilhelm (1883). *Einleitung in die Geisteswissenschaften: Versuch einer Grundlegung für das Studien der Gesellschaft und der Geschichte*. Leipzig: Duncker & Humblot.

Dilthey, Wilhelm (1900) [1976]. *Selected Writings*, edited and translated by H. P. Rickman, Cambridge: Cambridge University Press.

Dissanayake, Ellen (1992). *Homo Aestheticus: Where Art Comes from and Why*. New York: Free Press.

Dreyfus, Hubert (1992). *What Computers Still Can't Do: A Critique of Artificial Reason*. Cambridge, Massachusetts: MIT Press.

Eco, Umberto (1998). *Serendipities: Language and Luna*cy, translated by William Weaver. New York: Columbia University Press.

Edgar, Gerald (2004). *Classics on Fractals*. Boulder: Westview Press.

Edwards, Charles Henry (1994). *The Historical Development of the Calculus*. New York: Springer.

Einstein, Albert (1926). The Cause of the Formation of Meanders in the Courses of Rivers and of the So-called Baer's Law. *Die Naturwissenschaften*, Vol. 14.

Eisenlohr, A. (1877). *Ein mathematisches Handbuch der alten Aegypter (Papyrus Rhind des British Museum)*. Leipzig: J. C. Hinrich.

Elam, Kimberly (2001). *Geometry of Design: Studies in Proportion and Composition*. Princeton: Architectural Press.

Engelson, Morris (2017). The Biblical Value of Pi in Light of Traditional Judaism. *Journal of Humanistic Mathematics* 7: 37–71.

Euclid (1956). *The Thirteen Books of Euclid's Elements,* 3 Vols. New York: Dover.

Euler, Leonhard (1736). *Mechanica, sive motus scientia analytice exposita*. Petropoli: Typographia Academiae Scientiarum.

Euler, Leonhard (1748), *Introductio in analysin infinitorum*. Lausanne: Marcum-Michaelem Bousquet.

Eymard, Pierre and Lafon, Jean-Pierre (2004). *The Number Pi*. New York: American Mathematical Society.

Fauconnier, Gilles and Turner, Mark (2002). *The Way We Think: Conceptual Blending and the Mind's Hidden Complexities*. New York: Basic.

Fechner, Gustav (1876). *Vorschule der Ästhetik*. Leipzig: Breitkopf & Härtel.

Ferguson, Kitty (2008). *The Music of Pythagoras: How an Ancient Brotherhood Cracked the Code of the Universe and Lit the Path from Antiquity to Outer Space*. New York: Walker & Company.

Feynman, Richard Phillips and Robbins, Jeffrey (1999). *The Pleasure of Finding Things Out: The Best Short Works of Richard P. Feynman*. New York: Basic Books.

Fibonacci, Leonardo (1202) [2002]. *Liber Abaci*, trans. by L. E. Sigler. New York: Springer.

Flood, Ratmond and Wilson, Robin (2011). *The Great Mathematicians: Unravelling the Mysteries of the Universe*. London: Arcturus.

Fortnow, Lance (2013). *The Golden Ticket: P, NP, and the Search for the Impossible*. Princeton: Princeton University Press.

Fowler, D. H. (1987). *The Mathematics of Plato's Academy: A New Reconstruction*. Oxford: Clarendon Press.

Friedman, Tamar and Hagen, Carl R. (2015). Quantum Mechanical Derivation of the Wallis Formula for π. *Journal of Mathematical Physics* 56: https://doi.org/10.1063/1.4930800.

Friedman, Thomas L. (2007). *The World Is Flat: A Brief History of the Twenty-First Century*. New York: Picador.

Fritz, Kurt von (1945). The Discovery of Incommensurability by Hippasus of Metapontum. *The Annals of Mathematics* 46: 242–264.

Gardner, Howard (1983). *Frames of Mind: The Theory of Multiple Intelligences*. New York: Basic.

Gauss, Carl Friedrich (1966) [1801]. *Disquisitiones Arithmeticae*. New York: Springer.

Gernhardt, Ariane (2014). *Children's Drawings of Self and Family: Bridging Cultural and Universal Perspectives*. Dissertation. Universität Osnabrück.

Gillings, Richard J. (1961). Think-of-a-Number: Problems 28 and 29 of the Rhind Mathematical Papyrus. *The Mathematics Teacher* 54: 97–102.

Gillings, Richard J. (1962). Problems 1 to 6 of the Rhind Mathematical Papyrus. *The Mathematics Teacher* 55: 61–65.

Gillings, Richard J. (1972). *Mathematics in the Time of the Pharaohs*. Cambridge, MA: MIT Press.

Gödel, Kurt (1931). Über formal unentscheidbare Sätze der Principia Mathematica und verwandter Systeme, Teil I. *Monatshefte für Mathematik und Physik* 38: 173–189.

Grayson, Cecil (1972) *On Painting and On Sculpture: The Latin texts of De Pictura and De Statua*. London: Phaidon.

Gregory, James (1671/1938). *James Gregory: Tercentenary Memorial Volume, Containing His correspondence with John Collins and His Hitherto Unpublished Mathematical Manuscripts*. Edinburgh: the Royal Society of Edinburgh.

Grondin, Jean (1994). *Introduction to Philosophical Hermeneutics*. New Haven: Yale University Press.

Gupta, R. C. (1975). Madhava's and Other Medieval Indian Values of Pi. *Mathematics Education* 9 (3): 45–48.

Haag, Michael (2013). *Inferno Decoded: The Essential Companion to the Myths, Mysteries and Locations of Dan Brown's Inferno*. London: Profile Books.

Hall, Manly P. (1973). *The Secret Teachings of All Ages*. Los Angeles: Philosophical Research Society.

Hallyn, Fernand (1990). *The Poetic Structure of the World: Copernicus and Kepler*. New York: Zone Books.

Hardy, G. H. (1967). *A Mathematician's Apology*. Cambridge: Cambridge University Press.

Heath, Thomas L. (1958). *The Works of Archimedes with the Method of Archimedes*. New York: Dover.

Heidegger, Martin (2008). *Ontology: The Hermeneutics of Facticity*. Bloomington: Indiana University Press.

Herbert, Robert (1968). *Neo-Impressionism*. New York: The Guggenheim Foundation.

Hermann, Arnold (2005). *To Think Like God: Pythagoras and Parmenides—The Origins of Philosophy*. Las Vegas: Parmenides Publishing.

Hermite, Charles (1873). Sur la fonction exponentielle. *Comptes Rendus de l'Académie des Sciences de Paris* 77: 18–24.

Hersh, Reuben (1998). *What Is Mathematics, Really?* Oxford: Oxford University Press.

Hesiod (1987). *Theogony*, translated by R. S. Caldwell. Indianapolis: Hackett.

Hitchins, Derek (2010). *The Pyramid Builder's Handbook*. Morrisville: Lulu Enterprises.

Hofstadter, Douglas (1979). *Gödel, Escher, Bach: An Eternal Golden Braid*. New York: Basic Books.

Hofstadter, Douglas (1985). *Metamagical Themas*. New York: Basic Bookss.

Hofstadter, Douglas and Sander, Emmanuel (2013). *Surfaces and Essences: Analogy as the Fuel and Fire of Thinking*. New York: Basic.

Hodges, Andrew (1983). *Alan Turing: The Enigma*. New York: Vintage.

Homann-Wedeking, Ernst (1968). *The Art of Archaic Greece, Art of the World*. New York: Crown Publishers.

Hornung, Erik (1982). *Conceptions of God in Egypt: The One and the Many*. Ithaca: Cornell University Press.

Iosa, Marco, Morone, Giovanni, and Paolucci, Stefano (2018). Phi in Physiology, Psychology and Biomechanics: The Golden Ratio Between Myth and Science. *Biosystems* 165: 31–39.

Isacoff, Stuart (2003). *Temperament: How Music Became a Battleground for the Great Minds of Western Civilization*. New York: Knopf.

Ivins, William (1946). *Art and Geometry: A Study in Spatial Intuitions*. New York: Dover.

Izard, Véronique. Pica, Pierre, Spelke, Elizabeth S., and Dehaene, Stanislas (2011). Flexible Intuitions of Euclidean Geometry in an Amazonian Indigene Group. *PNAS* 108: 9782–9787.

James, Jamie (1995). *The Music of the Spheres: Music, Science, and the Natural Order of the Universe*. Göttingen: Copernicus.

Janson, Anthony F. (1995). *History of Art*. New York: Abrams.

Jones, Roger S. (1982). *Physics as Metaphor*. New York: New American Library.

Jones, William (1706). *Synopsis Palmariorum Matheseos, or a New Introduction to Mathematics*. London: J. Matthews.

Joost-Gaugier, Christiane L. (2006). *Measuring Heaven: Pythagoras and his Influence on Thought and Art in Antiquity and the Middle Ages*. Ithaca: Cornell University Press.

Jung, Carl G. (1972). *Synchronicity: An Acausal Connecting Principle*. Routledge and Kegan Paul.

Jung, Carl (2000), *Collected Works*. Princeton: Princeton University Press.

Jung, Carl G. (1983). *The Essential Jung*. Princeton: Princeton University Press.

Juster, Norton (1963). *The Dot and the Line: A Romance in Lower Mathematics*, by Norton. New York: Random House.

Kahn, Charles H. (2001). *Pythagoras and the Pythagoreans: A Brief History*. Indianapolis: Hackett Publishing Company.

Kangshan, Shen, Crossley, John N., and Lun, Anthony W. C. (1999). *The Nine Chapters on the Mathematical Art*. Oxford: Oxford University Press.

Kanigel, Robert (1992). *The Man Who Knew Infinity: A Life of the Genius Ramanujan*. New York: Washington Square Press.

Kaschube, Matthias, Schnabel, Michael, Löwel, Siegfrid, Coppola, David M., White, Leonard E., and Wolf, Fred (2010). Universality in the Evolution of Orientation Columns in the Visual Cortex. *Science* 330 (6007): 1113–1116.

Kasner, Edward and Newman, James (1940). *Mathematics and the Imagination*. New York: Simon and Schuster.

Katz, Mikhail G. and Sherry, David (2012). Leibniz's Infinitesimals: Their Fictionality, Their Modern Implementations, and Their Foes from Berkeley to Russell and Beyond. *Erkenntnis* 78: 571–625.

Keith, Michael (2010). *Not a Wake*. New York: Vinculum.

Klebanoff, Aaron (2001). π in the Mandelbrot Set. *World Scientific* 9 (4): 393–402.

Kepler, Johannes (1619). *Harmonices Mundi*. Lincii, Austriae: Planck.

Kline, Morris (1972). *Mathematical Thought from Ancient to Modern Times*, Vol. 1. New York: Oxford University Press.

Koestler, Arthur (1959). *The Sleepwalkers: A History of Man's Changing Vision of the Universe*. London: Hutchinson.

Kristeva, Julia (1980). *Desire in Language: A Semiotic Approach to Literature and Art*. New York: Columbia University Press.

Kulyukin, Vladimir A. (2007). Connect-the-Dots in a Graph and Buffon's Needle on a Chessboard: Two Problems in Assisted Navigation. *Semantic Scholar*. DOI:10.1142/9789812709677_0099.

Lakoff, George and Núñez, Rafael (2000). *Where Mathematics Comes From: How the Embodied Mind Brings Mathematics into Being*. New York: Basic Books.

Lambert, Johann Heinrich (1761). Mémoire sur quelques propriétés remarquables des quantités transcendentes circulaires et logarithmiques. *Histoire de l'Académie Royale des Sciences et des Belles-Lettres de Berlin* 17: 265–322.

Lévi-Strauss, Claude (1962). *La pensée sauvage*. Paris: Plon.

Lindemann, Ferdinand von (1882). Über die Zahl π. *Mathematische Annalen* 20: 213–225.

Liouville, Joseph (1844). Nouvelle démonstration d'un théorème sur les irrationnelles algébriques, inséré dans le Compte Rendu de la dernière séance. *Comptes Rendus de l'Académie des Sciences de Paris, Sér. A* 18: 910–911.

Listing, Johann Benedict (1848). *Vorstudien zur Topologie*. Göttingen: Vandenhoeck und Ruprecht.

Livio, Mario (2002). *The Golden Ratio: The Story of Phi, the World's Most Astonishing Number*. New York: Broadway Books.

Lobachevsky, Nikolai (1855). *Pangeometry*. Zürich: European Mathematical Society.

Lotman, Yuri (1991). *The Universe of the Mind: A Semiotic Theory of Culture*. Bloomington: Indiana University Press.

Macdonald, Fiona (2018). A Classic Formula For Pi Was Discovered Hidden in Hydrogen Atoms. *ScienceAlert*. https://www.sciencealert.com/formula-for-pi-has-been-discovered-hidden-in-hydrogen-atoms.

Mackenzie, Dana (2012). *The Universe in Zero Words*. London: Elwin Street Publications.

Malinowski, Bronislaw (1948). *Magic, Science, and Religion and Other Essays*. New York: Doubleday.

Mandelbrot, Benoit (1967). How Long Is the Coast of Britain? Statistical Self-Similarity and Fractional Dimension. *Science* 156 (3775): 636–638.

Mandelbrot, Benoit (1982). *The Fractal Geometry of Nature*. San Francisco: Freeman.

Mann, Tony (2010). From Sylvia Plath's *The Bell Jar* to the Bad Sex Award: A Partial Account of the Uses of Mathematics in Fiction. *BSHM Bulletin* 25: 58–66.

Maor, Eli (2007). *The Pythagorean Theorem: A 4,000-year History*. Princeton: Princeton University Press.

Marcus, Solomon (2012). Mathematics Between Semiosis and Cognition. In: Mariana Bockarova, Marcel Danesi, and Rafael Núñez (eds.), *Semiotic and Cognitive Science Essays on the Nature of Mathematics*, 98–182. Munich: Lincom Europa.

Martel, Yann (2001). *Life of Pi*. Toronto: Random House.

Martínez, Alberto (2012). *The Cult of Pythagoras: Math and Myths*. Pittsburgh: University of Pittsburgh Press.

Maturana, Humberto R. and Varela, Francisco (1973). *Autopoiesis and Cognition: The Realization of the Living*. Dordrecht: Reidel.

Mazur, Joseph (2008). *Zeno's Paradox: Unravelling the Ancient Mystery behind Space and Time*. New York: Plume.

Merton, Robert K. and Barber, Elinor (2003). *The Travels and Adventures of Serendipity: A Study in Sociological Semantics and the Sociology of Science*. Princeton: Princeton University Press.

Miller, Arthur I. (2001). *Einstein, Picasso: Space, Time, and the Beauty That Causes Havoc*. New York: Basic Books.

Mishlove, Jeffrey (1993). *The Roots of Consciousness*. New York: Marlowe & Company.

Mitchell, Kerry (2015). The Fractal Art Manifesto. https://www.fractalus.com/info/manifesto.htm.

Murtry, M. Ram and Murtry, V. Kumar (2013). *The Mathematical Legacy of Srinivasa Ramanujan*. New York: Springer.

Musser, Gary L., Burger, William F., and Peterson, Blake E. (2006). *Mathematics for Elementary Teachers: A Contemporary Approach*. Hoboken: John Wiley.

Myerson, Roger B. (1991). *Game Theory: Analysis of Conflict*. Cambridge: Harvard University Press.

Nassar, Sylvia (1998). *A Beautiful Mind*. New York: Simon & Schuster.

Newlands, John A. R. (1865). On the Law of Octaves. *Chemical News* 12: 83.

Nielsen, Michael (2012). *Reinventing Discovery: The New Era of Networked Science*. Princeton: Princeton University Press.

Nuessel, Frank (2013). The Representation of Mathematics in the Media. In: Mariana Bockarova, Marcel Danesi and Rafael Núñez (eds.). *Semiotic and Cognitive Science Essays on the Nature of Mathematics*, 154–198. Munich: Lincom Europa.

Oevermann, Ulrich, Allert, Tilman, Konau, Elisabeth, and Krambeck, Jürgen (1987). Structures of Meaning and Objective Hermeneutics. In: Volker Meja, Dieter Misgeld, and Nico Stehr (eds.), *Modern German Sociology: European Perspectives*, 436–447. New York: Columbia University Press.

Ogilvie, C. Stanley (1956). *Excursions in Mathematics*. New York: Dover.

O'Keefe, John and Nadel, Lynn (1978). *The Hippocampus as a Cognitive Map*. Oxford: Oxford University Press.

Olivastro, Dominic (1993). *Ancient Puzzles: Classic Brainteasers and Other Timeless Mathematical Games of the Last 10 Centuries*. New York: Bantam.

Oughtred, William (1631). *Clavis Mathematicae*. Lichfield: Veneunt Apud Tho. Robinson.

Pacioli, Luca (1509). *De divina proportione*. The Classic Us, 2013.

Palladio, Andrea (1570). *I quattro libri dell'architettura*. Venice: Marc Antonio Brogiollo.

Patras, Frédéric (2013). Mathématiques et herméneutique. *Archives de Philosophie* 76: 217–238.

Pearson, Karl (1905). The Problem of the Random Walk. *Nature* 72: 294.

Peet, Thomas E. (1923). *The Rhind Papyrus*. Liverpool: University of Liverpool Press.

Peirce, Charles S. (1923). *Chance, Love, and Logic*. New York: Harcourt, Brace.

Peirce, Charles S. (1931–1958) *Collected Papers of Charles Sanders Peirce*, Vols. 1–8, C. Hartshorne and P. Weiss (eds.). Cambridge, MA: Harvard University Press.

Pelkey, Jamin (2017). *The Semiotics of X: Chiasmus, Cognition, and Extreme Body Memory*. London: Bloomsbury.

Perkins, David (2000). *Archimedes' Bathtub: The Art and Logic of Breakthrough Thinking*. New York: W. W. Norton.

Petkovic, Miodrag S. (2009). *Famous Puzzles of Great Mathematicians*. Providence, RI: American Mathematical Society.

Petrie, William M. F. (1883). *The Pyramids and Temples of Gizeh*. London: Field & Tuer.

Pettigrew, Richard (2009). Aristotle on the Subject Matter of Geometry. *Phronesis* 54: 239–260.

Pickover, Clifford A. (2005). *A Passion for Mathematics*. Hoboken: John Wiley & Sons.

Plato (2008). *The Republic*. Project Gutenberg. https://www.gutenberg.org/files/1497/1497-h/1497-h.htm.

Plofker, K. (2009). *Mathematics in India*. Princeton: Princeton University Press.

Poincaré, Jules Henri (1890). Sur le problème des trois corps et les équations de la dynamique. Divergence des séries de M. Lindstedt. *Acta Mathematica* 13 (1–2): 1–270.

Poincaré, Jules Henri (1902). *Science and Hypothesis*. New York: Dover.

Polster, Burkard and Ross, Marty (2012). *Math Goes to the Movies*. Baltimore: Johns Hopkins University Press.

Pólya, George (1921). Über eine Aufgabe der Wahrscheinlichkeitsrechnung betreffend die Irrfahrt im Strassennetz. *Mathematische Annalen* 84: 149–160.

Pomeroy, Sarah B. (2013). *Pythagorean Women: The History and Writings*. Baltimore: The Johns Hopkins University Press.

Popkin, Richard H. (1966). *The Philosophy of the Sixteenth and Seventeenth Centuries*. New York: The Free Press.

Popper, Karl (1932). *The Logic of Scientific Discovery*. London: Routledge.

Popper, Karl (1963). *Conjectures and Refutations*. London: Routledge and Keagan Paul.

Popper, Karl (1972). *Objective Knowledge: An Evolutionary Approach*. Oxford: Oxford University Press.

Posamentier, Alfred S. (2004). *Pi: A Biography of the World's Most Mysterious Number*. New York: Prometheus.

Pratchett, Terry (2004). *Going Postal*. New York: Doubleday.

Raz, Amira, Packard, Mark G., Alexander, Gerianne M., Buhle, Jason T. Zhu, Hongtu, and Peterson, Bardley S. (2009). A Slice of π: An Exploratory Neuroimaging Study of Digit Encoding and Retrieval in a Superior Memorist. *Neurocase* 15: 361–372.

Read, John (1957). *From Alchemy to Chemistry*. New York: Dover.

Rescher, N. (2009). Reductio ad absurdum. *The Internet Encyclopedia of Philosophy*. https://www.iep.utm.edu/reductio/.

Rice, Michael (2003). *Egypt's Legacy: The Archetypes of Western Civilisation, 3000 to 30 B.C.* London: Routledge.

Richeson, David S. (2008). *Euler's Gem: The Polyhedron Formula and the Birth of Topology*. Princeton: Princeton University Press.

Riedweg, Christoph (2005). *Pythagoras: His Life, Teachings, and Influence*. Ithaca: Cornell University Press.

Roberts, Royston M. (1989). *Serendipity: Accidental Discoveries in Science*. New York: John Wiley.

Robinson, Abraham (1974). *Non-standard Analysis*. Princeton: Princeton University Press.

Rosenhouse, Jason and Taalman, Laura (2011). *Taking Sudoku Seriously*. Oxford: Oxford University Press.

Roy, Marina (2000). *Sign After the X*. Vancouver: Advance Artspeak.

Rucker, Rudy (2002). *Spaceland: A Novel of the Fourth Dimension*. New York: Tor.

Russell, Bertrand and Whitehead, Alfred N. (1913). *Principia Mathematica*. Cambridge: Cambridge University Press.

Sagan, Carl (1985). *Contact*. New York: Gallery Books.

Salmon, Wesley C. (ed.) (1970). *Zeno's Paradoxes*. Indianapolis: Bobbs-Merrill.

Schmidt, Bernd G. (ed.) (2000). *Einstein's Field Equations and Their Physical Implications: Selected Essays in Honour of Jürgen Ehlers*. New York: Springer.

Schneider, Michael S. (1994). *A Beginner's Guide to Constructing the Universe: The Mathematical Archetypes of Nature, Art, and Science*. New York: Harper Collins.

Schroeder, Lee (1974). Buffon's Needle Problem: An Exciting Application of Many Mathematical Concepts. *Mathematics Teacher* 67: 183–186.

Scott, Joseph F. (1981). *The Mathematical Work of John Wallis, D.D., F.R.S. (1616–1703)*. New York: Chelsea Publishing.

Sebeok, Thomas A. and Danesi, Marcel (2000). *The Forms of Meaning: Modeling Systems Theory and Semiotics*. Berlin: Mouton de Gruyter.

Sebeok, Thomas A. and Umiker-Sebeok, Jean (1983). You Know My Method. In: Umberto Eco and Thomas A. Sebeok (eds.), *Dupin, Holmes, Peirce: The Sign of Three*, 3–42. Bloomington: Indiana University Press.

Seebohm, Thomas M. (2007). *Hermeneutics: Method and Methodology*. New York: Springer.

Seskauskaite, Daiva (2013). The Significance of a Circle in Lithuanian Folklore and Mythology. *International Journal of Liberal Arts and Social Science* 1: 27–37.

Singh, Simon (1997). *Fermat's Enigma: The Quest to Solve the World's Greatest Mathematical Problem*. New York: Walker and Co.

Singh, Simon (2013). *The Simpsons and Their Mathematical Secrets*. New York: Bloomsbury.

Skiadas, Christos (ed.) (2016). *The Foundations of Chaos Revisited: From Poincaré to Recent Advancements*. New York: Springer.

Skinner, Stephen (2009). *Sacred Geometry: Deciphering the Code*. New York: Sterling.

Sklar, Jessica K. and Sklar, Elizabeth S. (eds.) (2012). *Mathematics in Popular Culture: Essays on Appearances in Film, Fiction, Games, Television and Other Media*. Jefferson: McFarland.

Stanciulescu, Traian D. (2014). Towards a Semiotics of sacred Geometry: On the Archetypal "architecture of Light." *IASS Publications*. DOI: 10.24308/iass-2014–117.

Stephenson, Rebecca (2011). Byrhtferth's Enchiridion: The Effectiveness of Hermeneutic Latin. In: Elizabeth M. Tyler (ed.), *Conceptualizing Multilingualism in England, c. 800–C. 1250*. Turnhout: Brepols.

Stewart, Ian (2001). *Flatterland: Like Flatland, Only More So*. New York: Basic Books.

Stewart, Ian (2008). *Taming the Infinite*. London: Quercus.

Stewart, Ian (2012). *In Pursuit of the Unknown: 17 Equations that Changed the World*. New York: Basic Books.

Stewart, Ian (2013). *Visions of Infinity*. New York: Basic Books.

Struik, Dirk J. (1987). *A Concise History of Mathematics*. New York: Dover.

Taylor, John (1859). *The Great Pyramid: Why Was It Built? And Who Built It?* Cambridge: Cambridge University Press.

Taylor, John R. (1995). *Linguistic Categorization: Prototypes in Linguistic Theory*. Oxford: Oxford University Press.

Thom, René (1975). *Structural Stability and Morphogenesis: An Outline of a General Theory of Models*. Reading: Benjamin.

Thom, René (2010). Mathematics. In: Thomas A. Sebeok and Marcel Danesi (eds.), *Encyclopedic Dictionary of Semiotics*, 3rd edition. Berlin: Mouton de Gruyter.

Thoreau, Henry D. (1864). *The Maine Woods*. Boston: Ticknor and Fields.

Turing, Alan (1936). On Computable Numbers with an Application to the Entscheidungs Problem. *Proceedings of the London Mathematical Society* 42: 230–265.

Turing, Alan (1952). The Chemical Basis of Morphogenesis. *Philosophical Transactions of the Royal Society of London. Series B, Biological Sciences* 237 (641): 37–72.

Twain, Mark (1889). *A Connecticut Yankee in King Arthur's Court*. New York: Harper and Brothers.

Uexküll, Jakob von (1909). *Umwelt und Innenwelt der Tierre*. Berlin: Springer.

Van Mersbergen, Audrey M. (1998). Rhetorical Prototypes in Architecture: Measuring the Acropolis with a Philosophical Polemic. *Communication Quarterly* 46: 194–213.

Venn, John (1888). *The Logic of Chance*. London: Macmillan.

Verene, Donald P. (1981). *Vico's Science of Imagination*. Ithaca: Cornell University Press.

Verner, Miroslav (2003). *The Pyramids: Their Archaeology and History*. London: Atlantic Books.

Villarroel, José Domingo and Ortega, Olga Sanz (2017). A Study Regarding the Spontaneous Use of Geometric Shapes in Young children's Drawings. *Educational Studies in Mathematics* 94: 85–95.

Vitruvius (1914). *Ten Books on Architecture*, trans. by M. H. Morgan. Cambridge: Harvard University Press.

Wallis, John (1655). *Arithmetica Infinitorum*. Leon: Lichfield.

Weiss, Piero and Taruskin, Richard (2008). *Music in the Western World: A History in Documents*. Cengage Learning.

Wells, David (1986). *The Penguin Dictionary of Curious and Interesting Numbers*. Middlesex: Penguin Books.

Wildgen, Wolfgang and Brandt, Per A. (2010). *Semiosis and Catastrophes: René Thom's Semiotic Heritage*. New York: Peter Lang.

Willerding, Margaret (1967). *Mathematical Concepts: A Historical Approach*. Boston: Prindle, Weber & Schmidt.

Wilson, John (2005). *The Official Razzie Movie Guide: Enjoying the Best of Hollywood's Worst*. New York: Grand Central Publishing.

Yee, Alexander J. (2019). Google Cloud Topples the Pi Record. http://www.numberworld.org/blogs/2019_3_14_pi_record/.

Zenil, Hector (ed.) (2013). *Irreducibility and Computational Equivalence*. New York: Springer.

Zimmermann, Jens (2015). *Hermeneutics: A Very Short Introduction*. Oxford: Oxford University Press.

Zöllner, Johan (1860). Ueber eine neue Art von Pseudoskopie und ihre Beziehungen zu den von Plateau und Oppel beschrieben Bewegugnsphaenomenen. *Annalen der Physik* 186: 500–525.

Zhu, Bokai, Dasco, Clifford C., and O'Malley, Bert W. (2018). Unveiling "Musica Universalis" of the Cell: A Brief History of Biological 12-Hour Rhythms. *Journal of the Endocrine Society* 2: 727–752.

Index